N. Hazon F.B. Eddy G. Flik · Ionic Regulation in Animals

Springer

Berlin
Heidelberg
New York
Barcelona
Budapest
Hong Kong
London
Milan
Paris
Santa Clara
Singapore
Tokyo

Neil Hazon F. Brian Eddy Gert Flik (Eds.)

Ionic Regulation in Animals: A Tribute to Professor W.T.W. Potts

With 80 Figures, 4 in Colour and 8 Tables

Springer

Dr. NEIL HAZON
Gatty Marine Laboratory
University of St. Andrews
St. Andrews, Fife, KY16 8LB
Scotland, U.K.

Dr. F. BRIAN EDDY
Department of Biological Sciences
University of Dundee
Dundee DD1 4HN
Scotland, U.K.

Dr. GERT FLIK
Department of Animal Physiology
University of Nijmegen
6525 ED Nijmegen
The Netherlands

ISBN-13:978-3-642-64396-5 Springer-Verlag Berlin Heidelberg New York

Library of Congress Cataloging-in-Publication Data

Ionic regulation in animals : a tribute to Professor W.T.W. Potts / N. Hazon, B. Eddy, G. Flik, editors.
p. cm.
Includes bibliographical references and index.
ISBN-13:978-3-642-64396-5 e-ISBN-13:978-3-642-60415-7
DOI: 10.1007/978-3-642-60415-7

1. Biological transport. 2. Ions--Physiological transport.
I. Potts, W. T. W. (William Taylor Windle) II. Hazon, N. (Neil)
III. Eddy, B. (Brian) IV. Flik, G. (Gert)
QH509.I68 1997 96-29511
572.6'41--dc21

Softcover reprint of the hardcover 1st edition 1997

Cover design: Design & Production, Heidelberg
Typesetting: Camera ready by Neil Hazon
SPIN 10535798 31/3137 5 4 3 2 1 0 - Printed on acid-free paper

Tribute

William Taylor Windle Potts was born in Sunderland, County Durham, 7 October 1928. His schooling was at the Perse School, Cambridge before taking a scholarship at St. Catharine's College, Cambridge and matriculating in 1949. His graduate work at Cambridge was under the renowned Prof. J.A. Ramsay, and the Ph.D. for studies of osmoregulation in molluscs was awarded in 1952. His first university appointment was as an Assistant Lecturer in Aberdeen (1952-54), followed by a Senior Lectureship at Birmingham from 1955-66. When the University of Lancaster was founded, he became its first Professor of Zoology in 1966 and was awarded the title Professor Emeritus in 1995.

Although Bill's primary interest has remained animal osmoregulation since his Cambridge days, his secondary interests in archaeology and British history have continued, most notably with the recent publication of the popular book Queen Victoria's Gene (1995, Alan Sutton) with his brother Malcolm. Bill's forte for the mathematical approach to basic physiological processes became apparent upon the publication of his often-referenced "The energetics of osmotic regulation in brackish and fresh water" in the J. Exp. Biol. in 1954. His publications in osmoregulation in aquatic animals included work on lamellibranch and cephalopod molluscs, and crustacea, with a single foray into the elasmobranch perilymph before the publication of his landmark book with Gwenyth Parry: Osmotic and Ionic Regulation in Animals in 1964. That volume summarized the field of animal osmoregulation in a way that would take an edited series today, established Potts and Parry as the leaders in the field, and provided the impetus and questions that led an entire generation of comparative physiologists into the study of how animals (not to mention cells) are able to maintain consistent body volume and composition in environments ranging from marine to freshwater and terrestrial. Two summers spent at the Marine Biological Laboratory at Woods Hole in the middle 1960s afforded Bill the opportunity to begin to study osmoregulation in fishes, and his publications since have alternated between this vertebrate taxon and invertebrate molluscs and arthropods.

Along with Jean Maetz in France, Bill was the first fish physiologist to routinely use radioisotopes to study fluxes across whole animals (e.g., Potts and Evans, 1967, Biol. Bull., 131: 362-368), and has since applied techniques such as measurement of transepithelial potentials (e.g., Potts and Eddy, 1973, J. Comp.

Physiol., 87: 29-48) and X-ray emission analysis (e.g., Potts and Oates, 1983, J. Exp. Zool., 227: 349-360) to characterize and quantify mechanisms of ionic transport. In recent years, he has developed an extensive collaboration with Russian scientists studying various aspects of the eco-physiology of the Aral Sea. His recent review of the kinetics of sodium uptake in freshwater animals (1994, Am. J. Physiol., 266: F315-320), 40 years after his first landmark paper (see above), pointed out, once again, the central position of his thinking in this field. This also has been shown by Bill's service as one of the Editors of the Journal of Comparative Physiology from 1976 to 1994.

Drs. Eddy, Flik, and Hazon brought together old friends and new colleagues of Bill Potts at the Lancaster meeting of the Society for Experimental Biology in late March 1996, to celebrate the effect that Bill Potts has had on our discipline and our own research. We dedicate these works to Bill Potts in celebration of his intellect, influence, and humanity.

David H. Evans
Professor of Zoology
University of Florida
Gainesville. FL 32611

Contents

Chapter 3 Regulating the micro-environment of ion transporting epithelia: A comparative approach

K.L. Shephard

Chapter 4 Osmotic and Ionic Osmoregulation in Cyclostomes

J.C. Rankin

Chapter 5 Ion and Water Balance in Elasmobranch Fish

N. Hazon, M.L. Tierney, W.G. Anderson, S. Mackenzie, C. Cutler and G. Cramb

Chapter 6 Gill Chloride Cells Activation by Plasma Osmolarity

J.A. Zadunaisky

Chapter 7 The use of Modern Microscopical Techniques for the Study of Fish Gill

A.J.H. van der Heijden and I.J. Morgan

Chapter 8 Transport and Housekeeping of calcium in Fish Gills

G. Flik

Chapter 9 Drinking in marine, euryhaline and freshwater teleost fish

J. Fuentes and F.B. Eddy

Chapter 10 Teleost Renal Function: Regulation by Arginine Vasotocin and by Angiotensins

J.A. Brown and R.J. Balment

Chapter 11 Arginine Vasotocin (AVT) Controls Renal Sodium and Water Excretion in Birds through Interaction with a new ADH Receptor Subtype

R. Gerstberger

Chapter 12 Intracellular signalling in salt-secreting cells – Recent Advances in the Avian Nasal Gland Model

T.J. Shuttleworth, J.L. Thompson and S.C. Martin

The Ecophysiology of Osmoregulation in Crustacea

R.R. Harris[1] and N.V. Aladin[2]

[1]Department of Zoology, University of Leicester, Leicester, LE1 7RH, U K

[2]Zoological Institute, Academy of Sciences, St Petersburg, 199034, Russia.

1.1 Introduction

Crustaceans are a remarkably successful group in both biomass and diversity terms. They occur in a wide range of aquatic environments ranging from open ocean to land - locked, transitory freshwater pools. A wide diversity of crustacean taxa have colonised estuaries, brackish and hypersaline lagoons and seas. Within many of these ecosystems they show significant adaptive radiation, particularly in the more geologically stable and large brackishwater seas. Here there is great crustacean diversity, and it is in the conditions prevalent in this type of ecosystem that the evolution of brackish and freshwater crustacean faunas probably occurred (Croghan, 1983; Barnes, 1989).

Much of their success as a group in this colonisation is due to their ability to tolerate, breed and develop in waters of different ionic concentration and composition. Different crustacean taxa possess osmoregulatory mechanisms which maintain their body fluid osmotic concentrations in steady - state, not only in media hypo-osmotic and hyperosmotic to their internal fluids (low salinity and high salinity) but also in rapidly fluctuating salinities. The group shows a wide variety of osmoregulatory strategies, some of which rely primarily on buffering changes in extracellular (haemolymph) osmoconcentration as the external salinity rises or falls, others involve intracellular adjustments of osmotic concentration, using specific osmotic effector molecules, in response to wide changes in extracellular concentration. Different osmoregulatory strategies have arisen as a result of natural selection and, within a single crustacean genus (even within a single species in the case of physiological races), wide differences in strategies now exist and are continuously subject to selective forces. Recent reviews have described many of the mechanisms underlying these strategies (Mantel & Farmer, 1983; Kirschner, 1991, Gilles & Pequeux, 1983).

The derivation of much of the crustacean fauna of northern European freshwater is probably from the ancient Sarmatic brackish seas (Sutcliffe, 1978; Croghan, 1983). In evolutionary history, colonisations and invasions by crustaceans have succeeded because of the invader's capacity to reproduce in the new habitats. This ability must have relied on the possession of osmoregulatory mechanisms with the capacity to take up ions from the low environmental ion concentrations characteristic of the Post Glacial freshwater bodies. Studies on populations of crustaceans inhabiting changing or marginal habitats have helped us understand how these colonisations may have occurred (Lockwood & Bolt, 1989). More recently, we have seen environmental changes resulting in increases in salinity in many natural waters. Man's activities have often lead to salination; often as a result of disposal of effluent from industrial or mining activites, which has created conditions suitable for invasive crustacean species originally from brackishwaters, often at the expense of native freshwater types. These alien species find the new water ionic concentrations within the tolerance limits of all life stages (e.g. Fryer, 1978, Van den Brink *et al.*,1991). However, in some cases excessive salination has resulted in the loss of endemic species. For example, the Aral Sea is presently undergoing such a process due to excess use of water for irrigation from inflowing rivers (Aladin, Plotnikov & Potts, 1995) Changes in the crustacean fauna of any water body subjected to salination will be the outcome of (a) loss of species due to the salinity increase outstripping the acclimatization capacities of existing species and (b) the persistence of physiological variants of the existing stocks (presumed to be genetically distinct) arising by natural selection. As to which of these two processes predominate must depend on the rate of change of environmental salinity.

This review will examine recent studies on the osmotic responses and tolerance of aquatic crustaceans concentrating on the response at the level of whole organism, with particular emphasis on environmental relations. Such differences in osmoregulatory behaviour between species, and of populations within species, will be examined. Recently, also, there has been a timely growth in interest in the osmoregulatory responses of the early stages in the life history of many crustaceans. A knowledge of the ontogeny of osmoregulation, and the way in which tolerance and steady - state patterns change from larvae to juveniles, and thence to adult, is clearly important in understanding the present distribution of crustaceans. Finally we shall examine how environmental change, either natural or anthropogenic in origin, can affect osmotic regulation and tolerance. Many dissolved compounds have been found to alter the exchange and uptake processes occurring across the crustacean integument. Where appropriate we shall refer to the way in which specific exchange process between the animals body fluids and the environment are changed by these factors.

1.2. Steady - state strategies

Much early work concentrated on examining salinity tolerance and patterns of extracellular steady - state following acclimation to different salinities. Three major patterns emerged out of these researches: hyperosmotic/isosomotic regulators, hyperosmotic/hypoosmotic regulators and conformers. Crustaceans exhibiting these responses can be either euryhaline or stenohaline. In addition freshwater species which hyperosmotically regulate their haemolymph within a relatively narrow external salinity range, have been defined as stenohaline. Most of these studies have been carried out on adult animals; much less is known of of the osmoregulatory capacities of young stages. In those conformers which show a wide variation in haemolymph osmotic concentration, yet are euryhaline, intracellular isosmotic regulation, involving the reduction or increase of intracellular solute concentrations must be occurring to maintain cellular integrity and function in the encountered salinity range (Mantel & Farmer, 1983; Kirschner, 1991, Gilles & Pequeux, 1983).

More recently, the term *amphiosmotic* has been introduced to describe that pattern in which haemolymph steady - state osmotic concentrations are maintained hyperosmotic at low salinities and hypoosmotic in higher ones (Aladin & Potts, 1995). In particular, Cladocera and Ostracoda have been looked at more widely and show interesting variations on the *amphiosmotic* theme. *Podonevadne angusta and Cornigerius maeoticus hircus* from the Caspian and Aral seas show a Type BI pattern in which haemolymph osmotic concentration is maintained at about 180 - 300 mosmol kg^{-1} over a salinity range from freshwater (FW) to 14 - 16 g l^{-1}. In the cladocerans, *Daphniaopsis pusilla* and *D. australis* from Australian salt lakes there is hyperregulation in low salinities, a wide range of near isosmoticity between haemolymph and medium (~270 to 670 mOsm kg^{-1}), and then a hyposmotic region above ~ 700 mOsm kg^{-1} (Type BII pattern). Ostracods from these geographical areas show these patterns also e.g. Type C1 *Leptocythere bacuana* (Caspian & Aral seas) and Type CII *Diacypris spinosa* a euryhaline species from Australia (Fig.1.1) Clearly many of these brackishwater forms have effective intracellular osmoregulatory mechanisms, with the capacity to tolerate wide salinity change (Aladin & Potts, 1996). Their ability to colonise hypersaline or changed salinity environments may result in them becoming more widespread in the future. Unique to this group, as far as is known, is the pattern of osmoregulation shown by the Caspian cladoceran *Penilia avirostris* . Thought to have evolved in freshwater, this species has invaded brackishwater but retains a low haemolymph osmotic concentration across a wide salinity range (~ 143 to 200 mOsm kg^{-1}). At the lower salinity limit, it is isosmotic with the environment (5 -7 g l^{-1}); above this it can tolerate field salinities up to ~ 40 g l^{-1}. The ostracod *Propontocypris maculata* also shows this pattern. Neither of these unusual steady - state osmoregulatory patterns have been described in other crustacean groups, to date, although there is considerable variation in the degree of hyperosmotic regulation i.e the haemolymph to medium difference, and the extent to which a constant plateau of osmotic concentration is maintained in different crustacean species.

Hyperosmotic regulation requires active ion uptake at a rate which can match the ion losses occuring across the integument and via renal organs. There is some evidence that ion uptake in the diet, or by drinking, are important routes for ion replacement in small crustaceans (Mantel & Farmer, 1983). Losses can be reduced by (a) lowering the integumental permeability to ions and water and (b) reducing the ionic concentration gradient between haemolymph and the external environment. Reduced osmotic permeability will slow the rate of water uptake and hence urinary flow. Since this is a potential route for ion loss this will be reduced also. In addition, reabsorption of ions in the antennary gland (or other renal organ), to produce a dilute urine, will further serve to reduce ion losses.

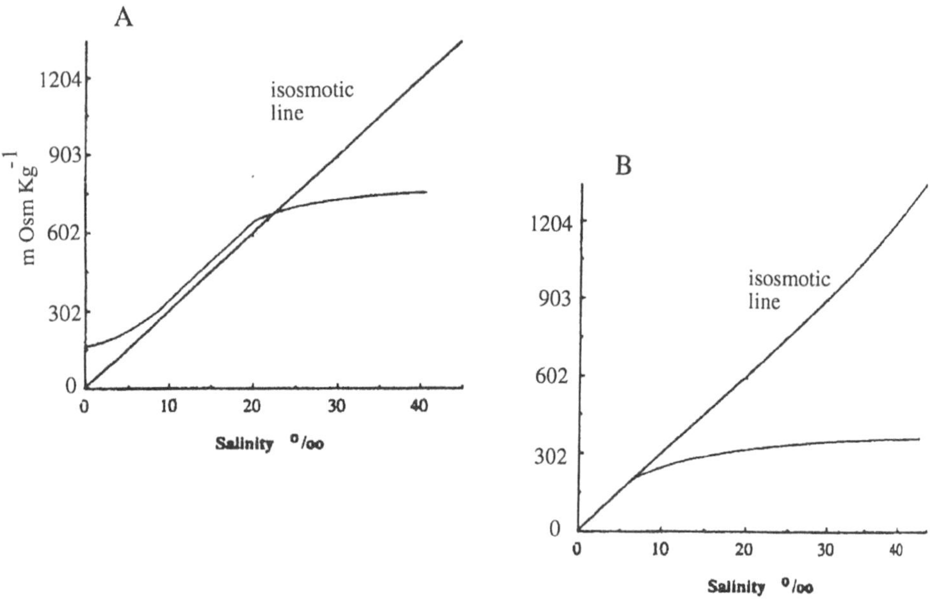

Fig. 1.1 Steady - state osmoregulatory patterns in Cladocera and Ostracoda. A. - *amphiosmotic* Type BII or CII showing a distinct isosmotic range. B - the pattern of *Penilia avirostris*, a hyposomotic regulator in salinities above the isosmotic point at 5-7 g l^{-1} salinity (Adapted from Aladin & Potts, 1995;1996).

Hypoosmotic regulation requires active ion extrusion to balance ion gains from (a) diffusional ion uptake from the higher ion concentration (hereafter [ion]) in the environment and (b) ion uptake with fluid imbibed from the saline external medium. This imbibition occurs to acquire the water to balance net osmotic losses to the environment. Ion uptake via both routes must be balanced by extrarenal mechanisms. For example, the neck organ of cladocerans and the mitochondrial - rich epithelial transporting cells (MRTC) in the inner shell layer of ostracods are regarded as ion extrusion organs, making possible the maintenance of the hypoosmotic steady - states described above (Potts & Durning, 1980; Aladin and Potts, 1996).

1.3 The development of osmoregulatory capacity

In his review of physiological adaptations to life in estuaries, Lockwood (1976) pointed out that many species, including crustaceans, have larval or juvenile stages which tolerate a reduced salinity range compared to adults of the species. The importance of this aspect of crustacean osmoregulation has often been underestimated when predicting the distribution of a species from tolerance studies.

It is clear that tolerance and steady - state patterns change significantly during development and that the extent of these changes is closely related to the ecology of a particular species. In the decapod *Penaeus japonicus* , for example, there is an improvement in low salinity tolerance at metamorphosis (here taken to be the end of the 3rd mysis stage) so that the 24 h lower lethal salinity (24 LC_{50}) increases from 25 to 7 g l^{-1} by post - larval stage 10 (Fig 1.2). In *Homarus americanus*, in contrast, early stages are more tolerant than *P. japonicus* to low salinity (LC_{50} = 17g l^{-1} at metamorphosis) but are less so in postlarval Stages IV to V (LC_{50} = 11g l^{-1}) (Charmantier *et al.*, 1988). In *P. japonicus* this increased tolerance to low salinities is correlated with the stage (post larval) when they migrate into coastal or estuarine waters whereas *H. americanus* has a lesser requirement for low salinity tolerance in its life history. Charmantier *et al.* (1988) proceeded to examine osmoregulation in both species and showed significant differences in steady - state patterns during development. In *P. japonicus*, early stages (6 nauplii, 3 zooea, 3 mysis) are slightly hyperosmotic (by 10 - 40 mOsm kg^{-1}) or conformers in the range 17 to 34 g l^{-1} but, after metamorphosis, there is a gradual shift to a hyperosmotic/hyposomotic pattern of regulation. In *H. americanus,* weak hyperosmotic/conformer regulation in stages I to III is replaced by a stronger hyperregulator pattern in stage IV, following metamorphosis. Similarly, in *H. gammarus,* this coincides with an increased tolerance to low salinities and is correlated with sharp rises between stage III and IV, in animals exposed to 17 g l^{-1} salinity, in the specific activity of the transport enzyme Na^{+}/K^{+}ATPase in the gills and pleopods (Thuet, 1988).

Species which breed in low salinities or freshwater must have free - swimming larval stages which show strong powers of osmoregulation. The atyid prawn *Macrobrachium petersi* has nine larval stages and breeds in FW where Stage 1

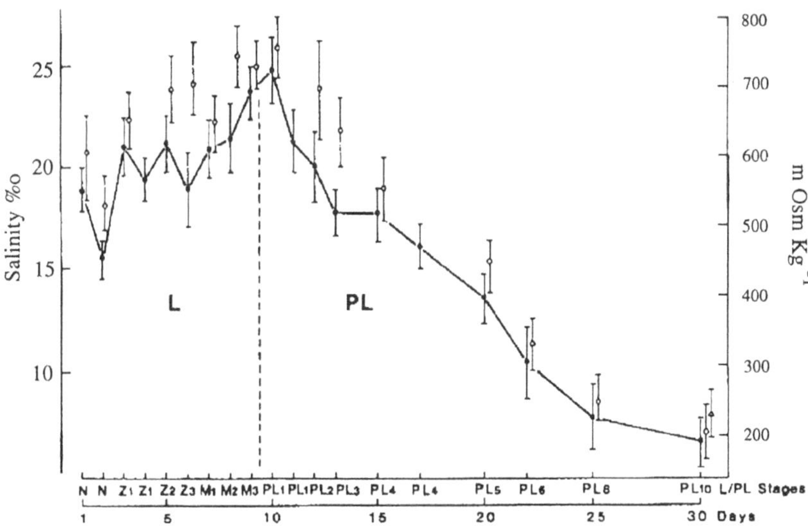

Fig. 1.2 Changes in low salinity tolerance during development of the shrimp *Penaeus japonicus* at 25°C. L = larval; PL = postlarval stages. The abscissa shows the different naupliar (N), zoeal (Z), and mysis (M) stages. Closed circles refer to 24 h LC_{50}, open circles to 48 h LC_{50} values. (Adapted from Charmantier *et al*, 1988).

larvae can maintain a hyperosmotic steady - state gradient of ~ 450 mOsm kg^{-1} between haemolymph and external medium, a level not dissimilar to the adult. However stage II to IX larvae lose the capacity to tolerate such low environmental [ion] and to regulate hyperosmotically. Interestingly, their ability to regulate hypoosmotically does not change, so that these stages can tolerate the brackishwater conditions found in the estuary (~ 5 - 40 g l^{-1}). Following metamorphosis, the capacity to maintain hyperosmotic steady state and tolerate low salinities is redeveloped, and post - larva and juveniles can move upstream into freshwater. In doing so they reduce their capacity to hypo-regulate as effectively as the larval stages, establishing the typical adult steady - state pattern (Read, 1984). Why osmoregulatory capacity is lost during intermediate stages is unclear. Presumably these larva can detect or avoid low salinities, or, if exposed, must tolerate temporary osmodilution until restored by currents or migration to brackish conditions. This loss of low salinity tolerance, would appear to be a risky strategy but must be of some significance for it to have been selected. It may reflect a change in energy partitioning, so that more energy is directed to organogenesis or growth at the expense of the requirement for osmoregulation. However, powerful ion pumps must be retained to generate the hypoosmotic steady - state in higher salinities. The capacity to re -direct ion transport to a hyperosmotic mode is seen in many other hyperosmotic/hypoosmotic regulators but, apparently, is not an option in this species.

Crustacean species that maintain populations in brackish waters, and do not migrate, develop low salinity tolerance and hyperosmotic regulation at an early stage. In the estuarine ghost shrimp *Callianassa jamaicense* var *louisianensis*, the short - lived, reduced larval stages (2 zooeal) are hyperosmotic regulators/ conformers, tolerating salinities as low as 3.5 g l^{-1} in ZI and 7 g l^{-1} in ZII stages. These salinities are within the range encountered in the burrows and habitats of the adults. Thus larval life history is entirely estuarine and the capacity to hyperosmotically regulate is shown at the earliest stage (Felder *et al.* 1986). Structural and biochemical changes necessary for maintainence of the steady - state appear early, some in the prehatch stages, whereas others can be observed in ZI, including high Na^+/K^+ATPase levels in the tissues and MRTCs in the epithelium of the inner branchiostegal surface, which show silver staining areas, often associated with ion - transport sites in the cuticle of crustaceans (Conte *et a*l, 1972). Thus specialised larval ion - transporting tissues are provided during development prior to the differentation of gills from the gill buds, the normal sites of ion uptake in adults. A feature of the ontogeny of osmoregulation in *Penaeus japonicus* is the appearance of special pleural areas which are also thought to be osmotic effectors (Bouricha *et al.* 1994).

The life history of *C. jamaicense* differs from that of other species of *Callianassa* in which larval stages develop in high salinity water offshore before settling in the lower reaches of estuaries. Many estuarine crustacean species show adult breeding migrations to offshore sites where larval stages can develop in high salinity conditions before returning to the adult habitats. This migration may be related to the availability of suitable feeding grounds for the larval stages. It would seem that, in these species, any larval osmoregulatory capacity over and above that required to cope with the immediate larval environment is not selected for. Also, that osmoregulatory tolerance at the cellular level is only developed in the larval stage in which it is required. Presumably, natural selection operates at the level of larval stages of different species resulting in a diversity of rates of development and different patterns of osmoregulatory capacity.

1.4 Brood pouches and osmoregulation

The role of brood pouches and *marsupia* in providing a suitable environment for developing crustaceans has received attention recently. Young stages can be protected from the osmotic changes experienced by the adult by being isolated to various degrees within the female. Following spawning in the isopod *Sphaeroma serratum*, eggs are transferred to fluid - filled deeply invaginating pouches formed in the ventral integument which open by a narrow slit to the outside. The fluid within is slightly hyperosmotic to the haemolymph of the female (~ 50 mOsm kg^{-1}) but considerably more so than the medium (~ 925 compared to 500 mOsm kg^{-1} in 17 g l^{-1} salinity medium). The body fluid of early embryos removed from the incubation pouch falls sharply when exposed directly to this salinity and they suffer fatal swelling and osmodilution within 10 - 15 hrs (Charmantier & Charmantier -Daures, 1994). It is clear that these early embryos require buffering from even moderately low salinities until the 1st juvenile stage when a greater degree of tolerance develops and juveniles leave the pouches and enter the *marsupium*. This structure is composed of oostegites and is of the "open" type through which circulation of the external medium occurs. Unlike the pouches, there is no difference in osmotic concentration betwen marsupial fluid and the environment. During juvenile Stages I to V, there is a progressive increase in the capacity to maintain a hyperosmotic haemolymph steady - state until the adult stage is attained.

In the brackish - water amphipod *Gammarus duebeni*, eggs are laid into an "open" marsupium where they are fertilised and eventually hatch into miniature adults. Adult *G. duebeni* tolerate a wide range of salinities and are hyperosmotic regulators but conformers below 21 g l^{-1} salinity (~ 600 mOsm kg^{-1}). The contrast with *S. serratum* is evident in that at no stage are developing embryos contained with a "closed" brood pouch, being always exposed to the environment in a well -ventilated oostegital *marsupium*. Not surprisingly, the capacity to hyperregulate body fluids appears early in the development of this species. The perivitelline fluid of Stage 2 and 3 embryos is highly hyperosmotic to the medium (~ 400 mOsm kg^{-1} in 2.6 g l^{-1} salinity ; ~75 mOsm kg^{-1}), and it is clear that all stages are able to tolerate a wide salinity range within or outside the *marsupium*. Unexpectedly, there is a switch to a hyperosmotic/hypoosmotic regulation pattern in Stages 5 - 7, followed by a restoration of the earlier pattern in hatchlings (Morritt & Spicer, 1995). These workers suggest that a more stable extracellular osmotic concentration may be required during certain critical stages of development. This aspect of the ontogeny of osmoregulation should be examined further since it may explain the changes in larval osmoregulatory pattern seen in other species.

In some species, however, the *marsupium* may have a role in protecting the early stages. The adult shore hopper *Orchestia gammarellus* is a euryhaline hyperosmotic/hypoosmotic regulator but its early embryos do not survive when directly exposed to ~ 9g l^{-1}. By Stage 3, however it can tolerate and grow in this salinity. In this species the marsupial fluid osmotic concentration is higher than that of the environment in the range 3.5 - 26 g l^{-1} as a result of the diversion of urine into the ventral groove. These early embryos are bathed in a milieu which is relatively stable in osmotic concentration (Morritt & Spicer, 1996). On the other hand, the survival rate of embryos of a freshwater amphipod *Gammarus pulex* out of the *marsupium* was greater in a weakly saline solution (2.5 g l^{-1} salinity) than in the normal low [ion] environment of the adult. This suggests that a degree of ion enrichment, sufficient for normal development, may occur in the *marsupium*, either by direction of urine or, by ion extrusion from MRTCs in the ventral plate region of some gammarids (Harris & Bayliss, unpublished observations).

In contrast, some estuarine mysids appear to remove ions from the marsupial fluid thus protecting developing embryos from full - strength sea water (McLusky & Heard, 1971).

Cladocera incubate eggs in brood chambers formed by the carapace, except when laying "resting eggs". In most brackish water and freshwater genera, e.g. *Daphnia, Sida* and *Simocephalus* , brood chambers are "open" and, presumably, well - ventilated so that the external medium osmotic concentrations bathing the eggs are similar to those of the environment. Those species with "closed" brood pouches, e.g. *Penila avirostris* from marine or brackish conditions, and *Polyphemus pediculus* from freshwater, maintain large osmotic gradients from inside to out, These gradients can be hyperosmotic or hypoosmotic depending on external conditions. Generally the embryos normally hatching in saline habitats are bathed by a fluid ~ 300 mOsm kg^{-1} lower than the external medium (~ 780 mOsm kg^{-1} in these experiments), while in freshwater, the fluid is ~ 240 mOsm kg^{-1} (Aladin & Potts, 1995). In both groups the "closed" brood pouch opens to the exterior late in development and embryos are exposed to the external osmotic conditions suddenly (Fig. 1.3). By this time effector organs have developed to ensure maintenance of the steady -state. These effector organs develop early in types with "open" brood pouches (Aladin, 1979; Halcrow, 1982).

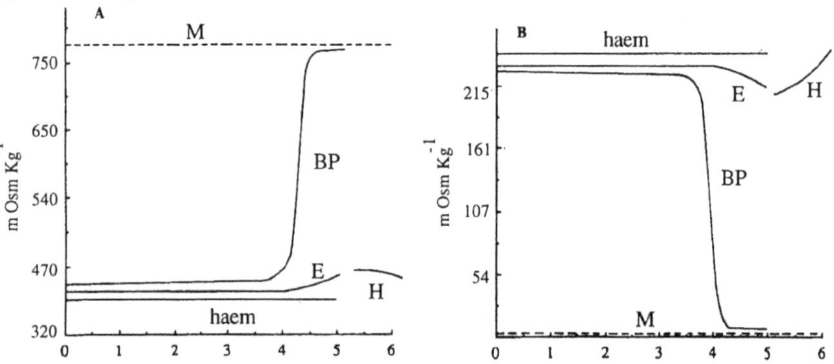

Fig. 1.3 Osmoregulation in the brood pouch and embryos of Cladocera with "closed" brood pouches. A - osmoregulation in saline conditions, B - osmoregulation in freshwater conditions. Ordinate - fluid osmotic concentrations (recalculated from Aladin & Potts, 1995); Abscissa - developmental stages 1-6. M = external medium, BP = brood pouch fluid, E = embryonic fluid, H = hatchling haemolymph, haem = adult haemolymph.

Direct development takes place in crayfish and potamonid crabs, important macrofauna in many freshwater ecosystems. Compression of the life cycle i.e. reduction or absence of larval stages, is often observed in freshwater and brackish water species . In these the vitelline membranes must play an important role is osmoregulation of the "imprisoned" larval stages even to the megalopa stage (Pace *et al*, 1976). On hatching, juveniles are exposed directly to the water conditions experienced by the adult, which will be ion - poor in many cases, although they may remain clinging to maternal pleopods temporarily. Surprisingly little is known of the ontogeny of osmoregulation in these groups. Morritt & Holdich (unpublished) report that in the crayfish *Pacifastacus* juveniles (Stage 1 hatchlings at < 24 h) have lower haemolymph osmotic concentrations, when hatched in freshwater, than adults (mean 378 compared to 430.5 mOsm kg^{-1}: latter value from Harris & Coley, 1991) and found that this species can be hatched successfully in 7 g l^{-1} salinity. Presumably these young individuals, being small, have a larger surface area to volume ratio and, therefore, high rates of ion turnover. They have to overcome many of the osmotic problems faced by adult amphipods and isopods which are of similar mass. Adult crayfish and freshwater crabs have relatively low ion and water permeabilities and probably expend less energy in maintaining steady - state.

In conclusion, various tolerances and patterns of osmoregulation have evolved in the development of crustaceans to enable larval life styles to be maintained in a great variety of habitats. More work needs to be done to elucidate how the development of capacity is regulated and how the selective forces, which may be acting on larval stages and adult stages alike, maintain this diversity. It is likely that intraspecific, as well as interspecific, differences in capacity exist between the larval stages of different crustacean populations. Individuals of one physiological race of a species may be more (or less) salinity tolerant than those of another race which has invaded, and become physiologically adapted, to a different habitat (see (**1.6**) below).

1.5 Environmental effects on crustacean osmoregulation

Consideration of the details of the physiological mechanisms of osmoregulation will be discussed elsewhere in this volume. We are concerned here with those environmental changes which may alter ion uptake and extrusion, or permeability. Recent models of ion uptake, mainly based on Na^+, support the idea of an external water - facing (apical) epithelial cell membrane Na^+ channel through which ions diffuse, at a rate dependent on a potential difference generated across that membrane by an active, electrogenic proton pump. (Ehrenfeld et al, 1985; Avella & Bornancin, 1989; Kirschner, 1988; Potts, 1994). This model can explain the apparent "saturation kinetics" of Na^+ uptake, observed in many aquatic animals, and the competitive inhibition of that uptake, previously interpreted by a "carrier" model involving Na^+/ H^+ or Na^+/ $NH4^+$ counterion exchange by an antiport carrier in the apical membrane. Both models require a baso-laterally located Na^+/K^+ATPase to maintain low intracellular [Na^+] for their operation. Since the models are current and have been used to explain various environmental effects on the ion uptake processes which are fundamental to the maintenance of steady

-state, it is important to try to explain each effect in terms of both models.

1.5.1 Effects of external [ion] or salinity

Environmental [ion] can be reduced by natural processes e.g. tidal water flows or increased freshwater run-off, to levels below the capacity of the ion uptake system of a particular species. If this occurs the rate of uptake is no longer adequate to balance the losses via the integument and renal systems and the crustacean will become depleted of ions. The measured uptake rates of Na^+ (also Cl^- and a few other ions) in a range of aquatic crustaceans has shown a correlation between "apparent ion affinity" (defined as K_m the external [ion] for half - maximal rate of uptake: m mol l^{-1}[ion]) and the normal environmental [ion]. Normally, freshwater species have low K_m (high affinity), and brackishwater/ marine species, high K_m (lower affinity) uptake systems (Kirschner, 1991 for review). The "capacity" of a particular uptake system i.e. the maximum rate of uptake in conditions where external [ion] are saturating (J_{max} or V_{max}), is often measured following ion - depletion to activate fully the uptake system. J_{max} has been taken to include an "emergency" capacity which is available in the short - term to replace ion losses if crustaceans become ion depleted e.g. tidal exposure to lower than normal salinity (Harris & Santos, 1993).

K_m and J_{max} are important osmoregulatory parameters which have arisen as a result of natural selection, and which adapt a species (or physiological race; see below) to particular environmental conditions. Kirschner (1991) has described the relationship between these uptake parameters and the rates of ion loss from an animal, which must be matched by a rate of uptake sufficient to achieve ionoregulatory steady - state. An example of this interrelationship is shown for the isopod *Sphaeroma serratum* (Fig. 1.4) (Harris & Thuet, 1987). Reductions in ion loss rate, for example, by lowering integumental ion permeability, would allow ion balance (here Na^+) to be achieved at a lower salinity. If K_m is low the sodium uptake system can generate a rate of uptake from a low salinity to balance losses. Clearly, a very steep curve of ion uptake (high affinity; lower K_m) could allow Na^+ to be acquired from the softest (ion -poor) of freshwaters. In the carrier - mediated model, K_m is a characteristic of the carrier affinity for Na^+, whereas J_{max} is a function of the density of carrier molecules in the apical membrane. In the channel - pump model, K_m must be dependent on the sodium conductivity of the apical membrane, or the relationship between outward proton pumping and the generated apical membrane potential difference. J_{max} would depend on the output of the proton pump (Potts, 1994). In both models, changes in K_m would require long -term changes in either intramembrane carrier proteins, or replacement of channel proteins. There is no evidence that short - term acclimation experiments can change K_m for Na^+ but, in some species, J_{max} has been altered by this means. For example, *Corophium curvispinum* , an invasive amphipod of brackishwater origins, exposed for 7 days to 0.5 m mol l^{-1} [Na^+] medium showed a significant increase in J_{max} for Na^+ but not for Cl^- (Bayliss & Harris, 1988, Harris & Bayliss,1990). Either the synthesis of more apical uptake sites, or

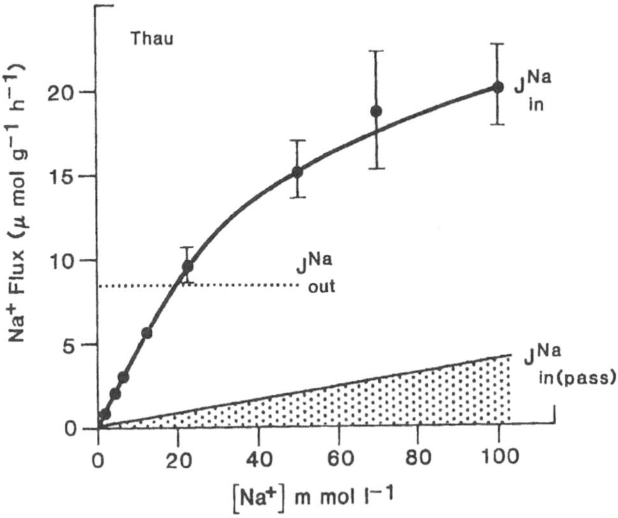

Fig. 1.4 The relationship between sodium influx (J_{in}) and external medium [Na^{+}] in *Sphaeroma serratum* (Isopoda). The active and passive ($J_{in(pass)}$) components of J_{in} are shown. The measured rate of efflux (loss) (J_{out}) is also shown. At the intersection of (J_{in}) and (J_{out}) Na^+ balance should be achieved Reprinted from J.Exp.Mar.Biol.Eiol. Vol.106, Harris and Thuet. "Physiological variability in geographically separated brackis-water and marine populations of *Sphaeorma* (Crustacea: Isopola Flabellifera) 279-297, 1987 with kind permission of from Elsevier Science - NL, Sara Burgerharttstraat 25, 1055 KV Amsterdam, The Netherlands

higher activity of baso - lateral Na^+/K^+ATPase, would be required to achieve increased J_{max} (Towle, 1990). In addition to interspecific differences in K_m and J_{max}, different populations within the same species have been found which have different values for these ion - uptake parameters.

There is clear evidence that interspecific differences in permeability, both to ions and to water, exist and that these differences are of adaptive significance permitting ion balance to be achieved in low [ion] environments (Kirschner, 1991 for review). More recently, changes in ion and water permeability have been shown to occur quite rapidly when crustaceans are transferred from one salinity to another. The amphipod *Gammarus duebeni*, following direct transfer from a low to a high salinity (0.7 - 35 g l^{-1}), temporarily reduces its rate of diffusional water exchange (measured as $t_{0.5}$ of ^{3}HHO exchange). This is low initially, decreases further, and then increases by 2.5 fold to the level normally found in seawater - acclimated animals (Lockwood & Bolt, 1989). Transfer of gammarids to higher salinities causes a temporary cessation of flow of blood through the coxal gills, the sites of ion and water exchanges (Taylor, 1985; Lockwood, 1993), These circulatory changes may alter the diffusional properties of the gill, in addition to any membrane changes that may occur (Spaargaren, 1989). Factors responsible for the characteristic interspecific (or population) differences in ion and water permeability may include the fatty acid composition of gill epithelial cell the characteristic interspecific (or population) differences in ion and water permeability may include the fatty acid composition of gill epithelial cell membranes (Morris *et al.*, 1987) and cuticular properties. Measurements of P_{Na}, the sodium permeability constant, of isolated gill cuticle gave values of 10^{-3}, 7.8 x 10^{-4} and 10^{-6} cm - sec^{-1} (lowest value = lowest permeability) in marine, brackishwater and freshwater decapods respectively (Lignon, 1987). Recent studies also suggest that ultrastructural features, particularly cytoskeletal microtubular bundles, help gill epithelial cells withstand the osmotic shocks of sudden salinity change, particularly at the apical regions of cell junctions. These septate junctions may also be adapted to cope with osmotic changes which may challenge the paracellular barrier between cells (Shires *et al.* 1994; Kikuchi & Matsumasa, 1993). *In vivo* reductions in ion and water permeability can reduce dissipative leaks and, therefore, are potentially energy conserving. They may be produced by modification of cuticlar properties and by changes in underlying cell membrane structure.

Crustaceans living in conditions of changing salinity, such as estuaries and intertidal areas, show many of the responses or adaptations described above. They are thus able to slow changes in their haemolymph or intracellular osmotic concentrations, or restore them by modifications of effector activity e. g. activate ion uptake systems. Environmental factors, which perturb these effector systems, can prevent the maintenance of steady - state or slow the rate of its restoration following salinity change.

1.5.2 Effects of water pH change on osmoregulation

Increases in $[H^+]$, predominantly a feature of freshwaters, occur in areas with acid soils and weak soil buffering capacity (Mason, 1989 for review). Water acidification in the presence of low $[Ca^{2+}]$ has been found to cause ion depletion in crustaceans (Vangenechten *et al*, 1989 for review), notably in crayfish (Wood & Rogano, 1986) and cladocerans (Potts & Fryer, 1979). In *Orconectes propinquus* 5 days exposure to pH 4.0 "soft" water produced severe decreases in haemolymph $[Na^+]$ and $[Cl^-]$ (by 25 and 20%, respectively). Similar observations have been recorded in other crayfish species (McMahon & Stuart, 1989; Jensen & Malte, 1990). In experiments involving longer - term exposure of *Procambarus clarkii*, a restoration of haemolymph $[Na^+]$ has been reported, implying some type of compensatory response to increased $[H^+]$ (Bennett & Walker, 1992). The initial inhibition of sodium influx following sudden exposure to high $[H^+]$, described by Shaw (1960) has been confirmed by many of these later studies.

In the" carrier" model, high external $[H^+]$ would reduce the rate of Na^+ / H^+ exchange and, if Na^+ efflux rate is maintained, a net Na^+ loss would occur resulting in haemolymph depletion. Cl^- influx is also reduced (resulting in Cl^- depletion), possibly by a lowering of internal $[HCO_3^-]$ and inhibition of the Cl^- / HCO_3^- counter - ion exchange across the gills (Wood & Rogano, 1986). A channel / proton pump mechanism would be affected by the raised external $[H^+]$ which would act to slow the rate of proton pumping. This, in turn, would reduce the potential difference driving force for sodium uptake. Therefore, both models show an interaction between environmental [Na^+] and [H^+] which resembles competitive inhibition for an apical uptake site, with increased [Na^+] tending to overcome the high level of influx inhibition at low medium [Na^+]. Additionally, reduced environmental pH appears to increase sodium permeability in some species. In the cladoceran *Daphnia magna*, sodium efflux increases by about 60% between pH 7 and 4, whereas *Acantholeberis curvisrostris* , a more acid resistant species, shows little change (Potts & Fryer, 1979). Sodium and chloride effluxes in the crayfish *Orconectes propinquus* were unaffected initially by pH 4 conditions but, with chronic exposure (5 days), sodium efflux gradually increased while influx showed some recovery (Wood & Rogano, 1986). There is some evidence, therefore, that Na^+ depletion by acid exposure results in activation of the sodium uptake system (Bennett & Walker, 1992). This could be, either by recruitment of additional carrier sites, or by a change in the relative affinities for Na^+ and H^+ of existing ones. In the channel/ proton pump model, this recovery of uptake would imply a change in the apical membrane H^+ permeability to restore net H^+ efflux rates and the potential difference driving force for Na^+ uptake (Potts, 1994).

1.5.3 Effects of ammonia and nitrite

Crustaceans, in common with freshwater fish, are susceptible to ammonia and nitrite when these compounds accumulate in the environment. This may be as a result of hypernutrification (excessive fertilizer use), or spillage of wastes containing nitrogenous compounds (e.g. agricultural waste). Excess levels of these pollutants outstrip the normal nitrification capacity of natural systems and toxic intermediates accumulate. Recently Foster & Turner (1993), using field experiments, showed the highly toxic effect of ammonia on crayfish. Ammonia, in addition to having cumulative toxic effects, has a significant effect on ionic regulation. Early studies showed that sodium influx in crayfish was inhibited markedly by low ammonia (Shaw, 1960), mainly in the form of ammonium ion, NH_4^+ (the pK of the NH_4^+/NH_3 equilibrium is 9.2 - 9.8, depending on ionic strength).

Our studies show that haemolymph [Na^+] in the crayfish *Pacifastacus leniusculus* decreased rapidly when exposed to freshwater medium containing 1.5 m mol l^{-1} total ammonia (T_{amm})(Fig. 1.5). This effect is most marked in high pH conditions when the ratio of NH_4^+ to NH_3 (unionised ammonia) was about 0.1 : 1.0. Under high T_{amm} conditions, haemolymph [Cl^-] showed no significant change. Sodium influx in Na^+ - depleted animals was more severly inhibited if NH_4^+ predominated (pH 8.2) rather than NH_3 (Fig. 1.6). Sodium influx measured over a range of environmental [Na^+], with and without ammonia (90% as NH_4^+), showed that sodium influx was inhibited by 67 - 75% at low [Na^+], but that, by increasing medium [Na^+], the inhibition of ammonia can be overcome. Further analysis needs to be done before we can state with confidence how ammonia acts to inhibit sodium influx. Armstrong *et al.* (1978) predicted, from ammonia toxicity studies on larvae of the euryhaline prawn *Macrobrachium rosenbergii*, that inhibition of sodium uptake was an important mode of toxic action of NH_4^+. Lin *et al* (1993) have confirmed that significant sodium and chloride depletion in juvenile *Penaeus japonicus* occurred following exposure to ammonia (16 - 32 mg l^{-1}: ~ 1.1 - 2.2 m mol^{-1} T_{amm}). Early life stages appear to be particularly susceptible to ammonia.

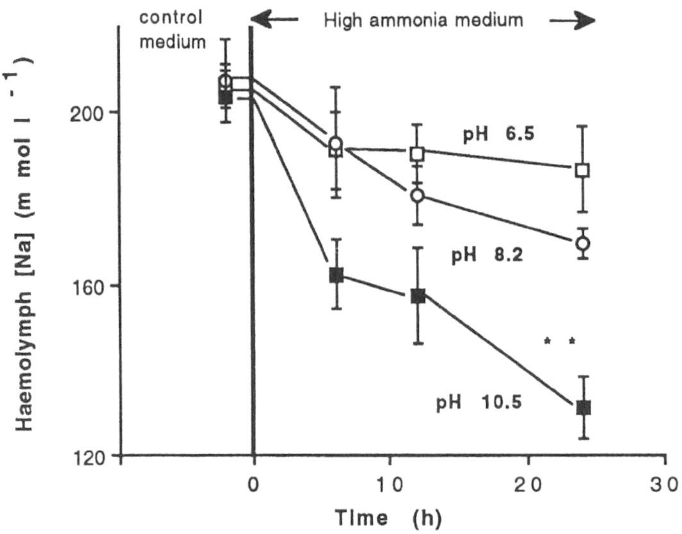

Fig. 1.5 The effect of external ammonia (T_{amm} = 1.5 m mol l^{-1}) on haemolymph [Na^+] in the crayfish *Pacifastacus leniusculus* in media of different pH. Calculated [NH_4^+] /[NH3] ratios for each medium are shown in Fig. 1.6. Means ± SEM (n=6).** indicates significantly different from all other means. (Harris & Coley, unpublished data).

In fish, the action of external ammonia, in the absence of significant Na^+ / NH_4^+ exchange, may be the result of an inward diffusion of NH_3 which binds intracellular H^+ causing an alkalisation in the cell and a reduced outward rate of proton pumping, the driving force for sodium influx (Avella & Bornancin, 1989). This may be the case in crustaceans also. However our results in crayfish (Harris & Coley, in preparation) show, that when inward diffusion of NH_3 should be greatest (and this must be the case judging from the rate of increase of haemolymph [T_{amm}]), the inhibition of sodium influx is relatively small (Fig. 1.6).

External nitrite (NO_2^-) is known to disrupt the chloride balance of freshwater crustaceans severely; its effects at higher salinities on brackish water species appears to be relatively minor (Jensen, 1990; Harris & Coley, 1991; Jeberg & Jensen, 1994; Jensen, 1995). Following exposure to 1.0 m mol l^{-1} [NO_2^-], chloride levels in the haemolymph fall sharply and NO_2^- is taken up against an electrochemical gradient and accumulated to high levels. NO_2^- competitively inhibits Cl^- uptake in crayfish (Harris & Coley, 1991) and is taken up across the gills by the active uptake system for that ion. Following accumulation in the haemolymph to high levels (to 25 m mol l^{-1} [NO_2^-]), nitrite starts to appear in the urine. Samples of urine taken from the nephropore show elevated [NO_2^-] and, unexpectedly, high [Cl^-]; normally a dilute urine is produced in crayfish (Fig. 1.7 a & b).This suggests that Cl^- uptake at the antennal gland is also inhibited when [NO_2^-] reaches a certain level, resulting in less chloride withdrawal and greater urinary Cl^- loss. This must contribute to the high net loss rate of Cl^- and, since NO_2^- is withdrawn in its place, reduces the rate of clearance of NO_2^- by the antennal gland increasing the rate of its accumulation in the haemolymph (Harris & Coley, unpublished; Jensen, 1995). The presently accepted model of chloride uptake remains that of a Cl^- /HCO_3^- counterion exchange (Potts, 1994) but, when nitrite is present externally, NO_2^- /HCO_3^- exchange does not appear to occur and haemolymph [HCO_3^-] levels rise (Harris & Coley, 1991).

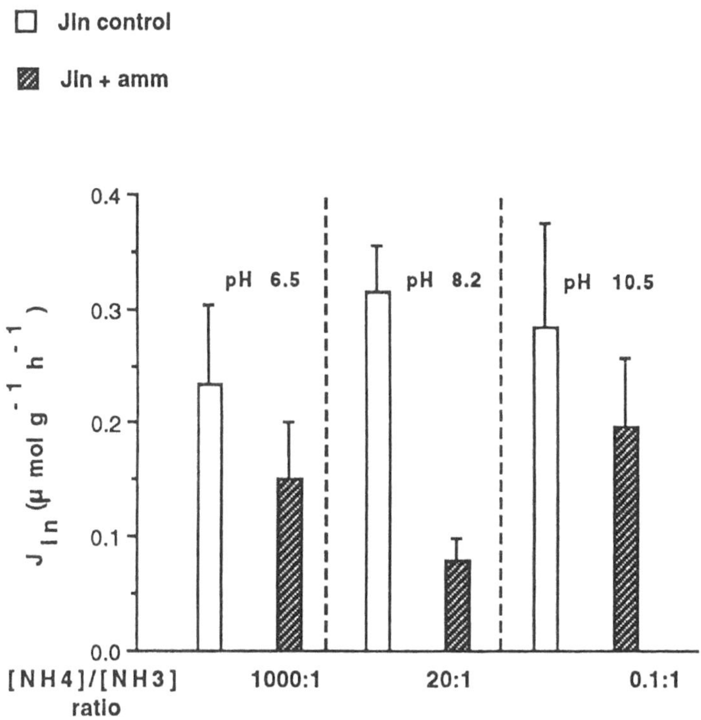

Fig.1.6 The effect of external ammonia (T_{amm} = 1.5 m mol l^{-1}) on Na^+ influx in the crayfish *Pacifastacus leniusculus*, Control and experimental values (+ amm) at each medium pH are shown together with calculated [NH_4^+] /[NH_3] on the abscissa. Means ± SEM (n = 6](Harris & Coley, unpublished data).

1.5.4 Other factors

Other environmental factors, natural or anthropogenic in origin, are known to affect crustacean osmoregulatory systems including dissolved heavy metals. Bambang *et al.* (1995a &1995b) have shown that sublethal concentrations of

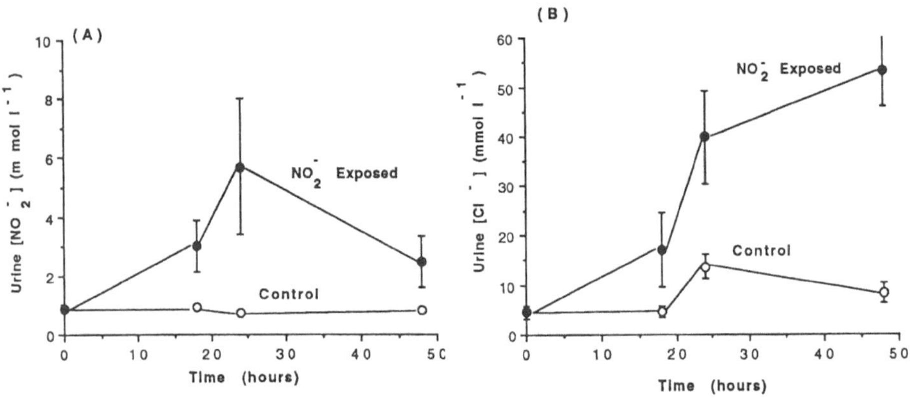

Fig.1.7 The effect of nitrite (NO_2^-) in the external medium on urine composition in the crayfish *Pacifastacus leniusculus*. A - urine [NO_2^-]; B - urine [Cl^-]. Control and nitrite exposed animals shown. Means ± SEM (n = 10-12).(Harris, unpublished data).

dissolved cadmium (Cd) and copper (Cu) reduces the capacity of larval stages of the shrimp *Penaeus japonicus* to maintain large hyperosmotic and hyposmotic haemolymph differences in low and high salinities, respectively. They suggest that damage to gill epithelial cells in the presence of Cd, and the inhibition of gill Na^+/K^+ATPase when Cu is present in the external medium, are responsible for reduced rates of ion uptake or extrusion, and that ion and water permeability changes may also occur. Further studies are required to show directly that impairment of ion transport is occurring during exposure to heavy metals but ultrastructural studies of MRTCs in the gills of*Carcinus maenas* show disruption of basolateral and apical membranes, and swelling of mitochondria, all of which may cause major changes in function (Lawson *et al*, 1995).

With the degradation of many of our natural waters increased levels of various toxicants may challenge the osmoregulatory processes taking place across the integument of crustaceans. Ultimately, the capacity of any species to tolerate particular salinity conditions lies in its ability to minimise the energy expenditure necessary to maintain osmoregulatory steady -state so that its cells can function normally. Challenges by toxicants will inevitably increase this energy expenditure, which may be particularly serious if the animal is exposed to extremes of salinity. The result will be that less energy is available for the growth and reproduction of the affected individual.

1.6 Intraspecific differences in osmoregulation

Studies of different populations of the same crustacean species have showed that within-species differences exist (comparison of different populations of the same species), and that these could allow individuals of such "physiological races" to invade a habitat quite different in salinity or [ion] from that of other populations. These populations are often given sub-specific rank because of slight morphological differences between them. Gammarids (Amphipoda) show such populations (Lockwood *et al* ,1976; Lockwood & Bolt, 1989; Taylor & Harris, 1986; Harris & Bayliss, 1990). *Gammarus duebeni celticucs*is found in freshwater in Ireland whereas *Gammarus duebeni duebeni* is regarded as a typical intertidal, estuarine sub-species capable of tolerating a wide salinity range, but not to the extent of living in freshwater (<0.5 g l^{-1} salinity), on the U. K. mainland, except for a few small, isolated populations on western coasts. If both sub - species are acclimated to very dilute seawater (0.25 m mol l^{-1} [Na^+], *G. duebeni celticus* maintains a higher haemolymph [Na^+] than *G. duebeni duebeni* (220 compared to 180 m mol l^{-1}). This significant difference in osmoregulatory capacity is the result of a number of different processes including, reduced ion permeability in freshwater, a lower apparent K_m for Na^+ uptake (0.4 - 0.7 compared to 1.5 - 2.4 m mol l^{-1}, respectively), and reduced diffusional permeability to water (Lockwood & Bolt, 1989). Changes in some of these parameters have been found in other species. In the case of *G. duebeni celticus*, the capacity to osmoregulate in freshwater is offset by an reduced ability to tolerate higher salinities. Presumably differences in the apparent K_m for Na^+ reflect differences in either the conductivity of the apical channels, or the relationship between the proton pumping rate and the driving force (the potential difference) generated by its activity; both of which could be subject to natural selection. In the carrier hypothesis, proteins with a higher or lower affinity for the substrate, Na^+ (or Cl^-), would be subject to selection, with the possible limitation that too high an affinity would increase the energy cost of separating carrier and substrate ion (Potts, 1994; Potts & Fryer, 1979).

Evidence that populations showing osmoregulatory differences can arise in a relatively short time - span, is shown in studies on the gammarid *Corophium curvispinum* ,which was originally described from the Caspian Sea and is presumed to be endemic in the Sarmatic Basin (Croghan, 1983). A freshwater variety *devium* (Wundsch) was reported in major European rivers in the 1930's and since then has spread extensively and rapidly in ion -rich freshwaters (Van den Brink *et al.*, 1991; Harris, 1991). This species may be considered a recent coloniser of freshwater, since some populations, living in the lowest [ion] environments, show reduced ion permeability and low K_m (Taylor & Harris, 1986), whereas others retain higher ion permeabilities (k, rate constant for Na^+ exchange, = 0.10 - 0.30 h^{-1}, compared to that of the estuarine species, *C. volutator*: = 0.50, and freshwater gammarids = 0.04 - 0.10 h^{-1})(Taylor, 1985b; Sutcliffe, 1975). *Corophium curvispinum* lives in environments with [Na^+] in the range 0.5 - 4.0 m mol l^{-1} [Na^+]. Some of the observed differences may be attributed to an acclimatization effect since laboratory acclimation (7 days), of individuals from one population to its lowest tolerable [Na^+], reduced k from

0.114 to 0.081 h^{-1}, and increased J_{max}. For this effect to be considered ecophysiologically significant, there should be minimal mortality in the test population showing the changed k and J_{max}. The survival of only a few individuals will mask real acclimatory change. Fig. 1.8 shows changes in *k* with time (1979 to 1995) at one site in the U. K., where water [Na^+]s were relatively high and showed large seasonal variations due to saline land run - off, and *k* values, measured recently, of animals from two Continental sites, the River Rhine and Lake Balaton . All sites are relatively ion - rich by normal freshwater standards. The populations living at the R. Rhine & Balaton sites, which have more stable [Na^+] (2.7 and 1.7 m mol l^{-1} [Na^+], respectively), have reduced values of *k* compared to the Grand Union population which retains a higher *k*, and, therefore, a higher rate of Na^+ turnover. Haemolymph [Na^+] values are similar in all these populations.

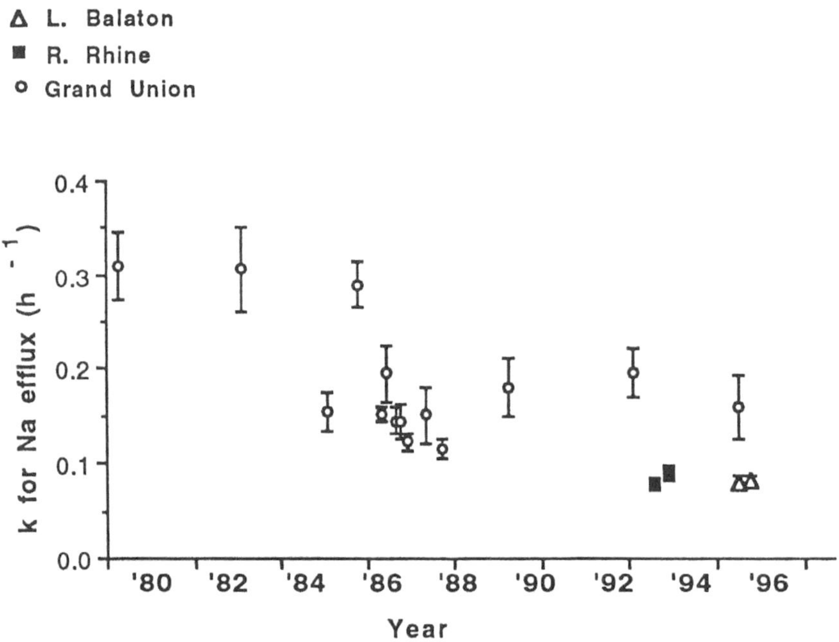

Fig. 1.8 Differences in sodium ion efflux (as rate constant, *k*: h^{-1}) of geographically separated populations of the amphipod *Corophium curvsipinum* Open circles show measurements of *k* made on a population from the Grand Union Canal, U.K. over the period 1979-1995. New measurements for R. Rhine and Lake Balaton animals are shown for 1993 and 1995, respectively. Means ± SEM are shown (n = 20 - 56, depending on site)(Harris & Andrews, unpublished data).

The aquatic isopods *Mesidotea* and *Sphaeroma* also show population (or racial) differences in osmoregulatory capacity. Baltic *Mesidotea entomon* and those from marine habitats in Arctic Canada do not tolerate freshwater, the latter having a lower limit of tolerance of ~ 1 g l^{-1} salinity (Percy, 1985). Freshwater *Mesidotea entomon* from both Swedish and Canadian lakes maintain high haemolymph osmotic concentration (~250 m mol l^{-1} NaCl or 530 mOsm kg^{-1}). This also has the lowest water permeability compared to marine populations (k = 0.15 compared 0.35 h^{-1}). *M. entomon* in Swedish ion - poor lakes ($[Na^+]$ = 0.26 m mol l^{-1}) show lower sodium permeability and K_m for Na^+ uptake than those from the Baltic, or from those from L. Malaren, which was formed only recently (~ 700 years ago) from an arm of the Baltic. A combination of increased uptake rate and reduced loss rate allows these freshwater races to maintain osmoregulatory steady - state in low [ion] water (Lockwood & Bolt, 1989). Under circumstances where [ion] is changing gradually, here becoming more freshwater, selection must occur towards the adapted osmoregulatory genotype in an isolated population.

Cladocera, with their short generation time and rapid rate of reproduction, can colonise new water - bodies and diverge rapidly both genetically and phenotypically from parent stocks . *Daphnia magna* occurs in snow melt water lakes and ponds in the Canadian tundra. Some pools are very ion - poor, whereas others are influenced by sea spray, so that a range of salinity conditions are found. High salinity tolerance of juvenile and adults from the ion - poor ponds was less than that of clones from seaward pools ($[Na^+]$ up to 166 m mol l^{-1}), whose populations could reproduce in saline conditions. From enzyme polymorphism studies, it was found that these populations showed a high degree of genetic differentiation. Large differences in osmoregulatory capacity also appear to have arisen within a short time span. Clones from ion-poor conditions maintain haemolymph osmotic concentrations two - fold higher than the more salinity tolerant ones, when tested in the same salinity, suggesting a change from a strongly regulating to a more conforming phenotype, within the same species (Weider & Hebert, 1987). Such large differences in osmoregulatory strategy *within the same species* have not been reported as yet in other crustacean groups. There is evidence that in *D. magna* , K_m and J_{max} differ in some U. K. populations (apparent K_m ranges from 0.4 to 0.05 m mol l^{-1} $[Na^+]$: Potts & Fryer, 1979; Stobbart *et al*, 1977). However, physiological differences in such isolated populations may arise a result of genetic drift, and may not be in the expected (adaptational) direction.

In some cases, also, the correlation between ecology and osmoregulatory capacity is not immediately evident. The lower sodium permeability (P_{Na}) of *Sphaeroma serratum* (Isopoda) population from the Etang (lagoon) de Thau, S. France, where it experiences only slight salinity dilution, can be contrasted with that of one from the Etang de Vic which has a more variable salinity regime, often falling to elevels where the species hyper-regulates (P_{Na}: Thau = 1.98 . 10^{-8}; Vic = 4.35 . 10^{-8} m s^{-1}; Harris & Thuet, 1987). However, the Vic population generates > two-fold more Na^+ uptake at low salinity levels (K_m = 48.5 (Vic) and 41.9 (Thau) m mol l^{-1} $[Na^+]$), and can achieve sodium balance in 19.5 m mol l^{-1} $[Na^+]$, a values similar to the balance point of Thau animals (Fig. 1.4). This population may be expected to expend more energy in Na uptake during those period when salinity is low. However some reduction in haemolymph $[Na^+]$ to levels lower than other *Sphaeroma populations* may occur under these conditions.

More detail of the pattern of haemolymph [Na^+] in low salinities is required to resolve this point.

1.7 Conclusions

Various capacities and patterns of osmoregulation have evolved within the Crustacea resulting in wide interspecific differences in ability to cope with different aquatic habitats. In some cases we can see how selection can operate to produce different osmoregulatory strategies. In all groups a combination of factors, including ion affinity, uptake capacity and permeability, interact to achieve steady - state in a particular environmental salinity. Presumably there are additional energy costs in sustaining high ion uptake rates. These could be lower if ion permeability was reduced, thus effecting a decreased rate of ion turnover. In the populations with high rates of ion turnover above, the energy requirement may comprise a large proportion of the energy budget of the animal. At the margins of the salinity tolerance range of any species, these costs may reduce, in the short - term at least, its scope - for - growth (Bayne et al, 1979) i.e. that proportion of its budget available for growth and reproduction. This has yet to be demonstrated in terms, say, of the occurrence of smaller - sized, slow - growing animals, and reduced fecundity, in any crustacean population which could be attributed to osmotic stress, to our knowledge. In isolated populations selective pressure may operate on larval or juvenile stages reducing the diversity of osmoregulatory capacities within the population. These could become re-established in a population if selective pressures were to be relaxed, for example, as a result of a change in water quality.

Intuitively, one would ascribe differences in the osmoregulatory capacities of isolated populations as being due to genotypic differences. What may also become evident, with further study, is that intraspecific, as well as the observed interspecific, differences in salinity tolerance and osmotic steady - state capacity exist at the level of the larval and juvenile stages. There is an urgent need for the establishment of breeding cultures and the technology for the rearing of successive generations under specified salinity conditions, with appropriate cross - breeding experiments, in order that we can demonstrate the heritability of important osmoregulatory parameters such as ion and water permeability, and ion uptake capacity and affinity. It will be necessary also to establish the limits of phenotypic osmoregulatory adaptability, or acclimation effects, within different species. There have been very few studies of such long term changes within either natural or cultured crustaceans species (e.g.Sutcliffe, 1971).Aquatic crustaceans would appear to possess many useful features for genetical studies such as robustness, relatively short generation times and, therefore, many generations per year.

Evaporation of water from many of the Earth's water bodies may become a feature of global warming, and this will result also in salination, at least in some geographical areas (Clayton, 1995). Existing freshwater and brackishwater crustacean faunas will become subjected to osmoregulatory and ionoregulatory stresses. Increases in sea level will cause the inundation of coastal areas, which are often fringed by brackish lagoons and mangrove swamps. Their salination, either by inundation or by subterranean seepage, will also cause profound faunal changes among crustaceans which are often the dominant macroinvertebrates in such ecosystems. Freshwater lakes, adjacent to coastal areas, will also be subject to increased salinity. Any faunal changes that follow will involve both loss of species and invasion by species which have a more marine or euhaline distribution at present. It will be important to be able to predict some of these ecological changes and a knowledge of the salinity tolerance and osmoregulatory capacity of existing species, or races within species, will help in this respect.

Finally, Man has used his knowledge of the osmoregulatory capacities of certain species to select suitable stocks for crustacean aquaculture and, in many cases, created salinity conditions for the culture of a particular species. These developments e. g. the impounding of coastal areas to create lagoons in mangrove areas, will cause profound changes to the salinity regimes of existing ecosystems and the displacement of natural faunas.

References

Aladin, N.V. (1979) Morphophysiological adaptation of marine Cladocera (in Russian). Autoref.kand. diss. L.;38.

Aladin, N.V., Plotnikov I.S. and Potts, WTW (1995) The Aral sea dessication and possible ways of rehabiliting and conserving its northern part. Envirometrics 6: 17-29.

Aladin, N.V. and Potts, W.T.W. (1995) Osmoregulatory capacity of the Cladocera. J Comp Physiol B 164: 671-683.

Aladin, N.V. and Potts W.T.W. (1996) Osmoregulatory capacity of the Ostracoda. J Comp Physiol B 166:

Armstrong, D.A., Chippendale, D., Knight, A.W. and Colt, J.E. (1978) Interaction of ionised and un-ionised ammonia on short-term survival and growth of prawn larvae *Macrobrachium rosenbergii*. Biol Bull 154: 15-31.

Avella, M. and Bornancin, M. (1989) A new analyisis of ammonia and sodium transport through the gills of the freshwater rainbow trout (*Salmo gairdneri*). J Exp Biol 142: 155-175.

Bambang, Y., Charmantier, G., Thuet, P. and Trilles, J-P. (1995b) Effect of cadmium on survival and osmoregulation of various developmental stages of the shrimp *Penaeus japonicus* (Crustacea: Decapoda). Mar Biol 123: 443-450.

Bambang, Y., Thuet, P., Charmantier-Daures, M., Trilles, J-P. and Charmantier, G. (1995a) Effect of copper on survival and osmoregulation of various developmental stages of the shrimp *Penaeus japonicus* Bate (Crustacea: Decapoda). Aquatic Toxicol 33: 125-139.

Barnes, R.S.K. (1989) What, if anything, is a brackish water fauna? Trans Roy Soc Edinburgh: Earth Sciences 80: 235-240.

Bayliss, D. and Harris, R.R. (1988) Chloride ion regulation in the freshwater amphipod *Corophium curvispinum* and acclimatory effects of external Cl^-. J. Comp Physiol B 158:81-90.

Bayne, B.L., Moore, M.N., Widdows. J., Livingston, D.R. and Salkeld. P. (1979) Measurement of the responses of individuals to environmental stress and pollution: studies with bivalve molluscs. Phil Trans Roy Soc Lond B 286: 563-581

Bennett KM, Walker RL (1992) Effects of acid exposure on acid-base, electrolyte status and gill Na/K-ATPase activity in crayfish (*Procambarus clarkii*) Am Zool 32: 47A

Bouaricha, N., Charmantier-Daures. M., Thuet, P., Trilles, J-P. and Charmantier G (1994) Ontogeny of osmoregulatory structures in the shrimp *Penaeus japonicus* (Crustacea; Decapoda). Biol Bull 186: 29-40.

Charmantier. G., Charmantier-Daures, M., Bouaricha, N., Thuet, P., Aiken, D.E. and Trilles J-P (1988) Ontogeny of osmoregulation and salinity tolerance in two decapod crustaceans: *Homarus americanus* and *Penaeus japonicus*. Biol Bull 175: 102-110

Charmantier, G, and Charmantier-Daures, M. (1994) Ontogeny of osmoregulation and salinity tolerance in the isopod *Sphaeroma serratum*. Mar Ecol Prog Ser 114: 93-102.

Clayton, K. (1995) The threat of global warming. In: O'Riordan T (Ed) Environmental Science for Environmental Management. Longman Scientific. pp 110-130

Conte, F., Hootman, S. and Harris P (1972) Neck organ of *Artemia salina* nauplii: a larval salt gland. J Comp Physiol B 80: 239-246.

Croghan PC (1983) Osmotic regulation in the evolution of brackish and fresh-water faunas. J Geol Soc 140: 39-46.

Ehrenfeld, J., Garcia-Romeu, F. and Harvey BJ (1985) Electrogenic active proton pump in *Rana esculenta* skin and its role in sodium transport. J Physiol Lond 359: 331-355.

Felder. J., Felder, D. and Hand, S.C. (1987) Ontogeny of osmoregulation in the estuarine ghost shrimp *Callianassa jamaicense* var *Louisianensis* Schmitt (Decapoda: Thalassinidae). J Exp Mar Biol Ecol 99:91-105.

Foster. J. and Turner, C. (1993) Toxicity of field-simulated farm waste episodes to the crayfish *Austropotamobius pallipes*(Lereboullet): elevated ammonia and reduced dissolved oxygen. Freshwater Crayfish 9: 249-258.

Fryer, G. (1978) A remarkable inland brackish-water crustacean fauna from the lower Aire valley, Yorkshire. A conundrum for the ecologist. Naturalist 103: 83-94.

Gilles, R. and Pequeux, A, (1983) Interactions of chemical and osmotic regulation with the environment. In: Bliss DE (Ed) The Biology of the Crustacea. Vol 8. Academic Press, pp 109-177.

Halcrow, K. (1982) Some ultrastructural features of the nuchal organ of *Daphnia magna* Straus Crustacea: Branchipoda. Can J Zool 63: 738-740.

Harris, R.R. (1991) Amphipod also invades Britain. Nature 354: 194.

Harris, R.R. and Bayliss, D. (1990) Osmoregulation in *Corophium curvispinum* (Crustacea: Amphipoda), a recent coloniser of freshwater. III Adaptive changes in sodium regulation. J Comp Physiol B 160: 85-92.

Harris, R.R. and Coley, S. (1991) The effects of nitrite on chloride regulation in the crayfish *Pacifastacus leniusculus* Dana (Crustacea: Decapoda). J Comp Physiol B 161: 199-206.

Harris, R.R. and Santos, M.C.F. (1993) Sodium uptake and transport (Na^{+}+K^{+})ATPase changes follwing Na^{+} depletion and low salinity acclimation in the mangrove crab *Ucides cordatus* (L.) Comp Biochem Physiol 105A: 35-42.
Harris, R.R. and Thuet, P. (1987) Physiological variability in geographically separated brackish-water and marine populations of *Sphaeroma* (Crustacea: Isopoda: Flabellifera). J Exp Mar Biol Ecol 106: 279-297.
Jeberg, M.V. and Jensen, F.B. (1994) Extracellular and intracellular ionic changes in crayfish *Astacus astacus* exposed to nitrite at two acclimation temperatures. Aquatic Toxicol 29: 65-72.
Jensen, F.B. (1990) Sublethal physiological changes in freshwater crayfish *Astacus astacus*, exposed to nitrite: haemolymph and muscle tissue electrolyte status and haemolymph acid-base balance and gas transport. Aquatic Toxicol 18: 51-60.
Jensen, F.B. (1995) Uptake and effects of nitrite and nitrate in animals. In: Walsh PJ, Wright P (Ed) Nitrogen metabolism and excretion. CRC Press, pp 289-303
Jensen, F.B. and Malte, H. (1990) Acid-base and electrolyte regulation, and haemolymph gas transport in crayfish *Astacus astacus,* exposed to soft acid water with and without aluminium. J Comp Physiol B 160, 483-490
Kikuchi, S. and Matsumasa, M. (1993) The osmoregulatory tissue around the afferent blood vessels of the coxal gills in the estuarine amphipod *Grandidierella japonica* and *Melita setfifgella* Tissue Cell 25: 627-638.
Kirschner, L.B. (1988) Basis for apparent saturation kinetics of Na^{+} in freshwater hyperregulation. Am J Physiol 254 (Regulatory integrative Comparative physiology 23): R984-R988.
Kirschner, L.B. (1991) Water and ions. In Ladd Prosser C (Ed) Comparative Animal Physiology 4th Edition Wiley-Liss, pp 13-107
Lawson, S.L., Jones, M.B. and Moate, R.M. (1995) Effect of copper on the ultrastructure of the gill epithelium of *Carcinus maenas* (Decapoda: Brachyura). Mar Poll Bull 31: 63-72.
Lignon, J.M. (1987) Ionic permeabilities of the isolated gill cuticle of the shore crab *Carcinus maenas* J Exp Biol 131: 159-174.
Lin, H-P., Thuet, P., Trilles, J-P., Maunet-Guillaume, R. and Charmantier, G. (1993) Effect of ammonia on survival and osmoregulation of various developmental stages of the shrimp *Penaeus japonicus*. Mar Biol 117: 591-598
Lockwood, A.P.M. (1976) Physiological adaptation to life in estuaries. In: Newell RC (Ed) Adaptation to Environment - essays in the physiology of marine animals. Butterworths.
Lockwood, A.P.M. (1993) *Gammarus duebeni* : a flexible phenotype. In: Aldrich JC (Ed) Quantified phenotypic responses in morphology and physiology. Proc. 27th Eur Mar Biol Symp, pp 215-227.
Lockwood, A.P.M. and Bolt, S.R.L. (1989) Physiology of crustacea from difficult environments. Trans Roy Soc Edinburgh (Earth Sciences) 80: 285-292

Lockwood, A.P.L., Croghan. P.C. and Sutcliffe, D.W. (1976) Sodium regulation and adaptation to freshwater in Crustacea as examplified by the isopod *Mesidotea entomon* and the amphipod *Gammarus duebeni*. In: Spencer Davies P (Ed) Perspectives in Experimental Biology Vol 1 Zoology, Pergamon Press, pp 93-106

Mantel, L.H. and Farmer, L.L. (1983) Osmotic and ionic regulation. In: Bliss DE (Ed) The Biology of the Crustacea Vol 5, Academic Press., pp 54-161.

Mason, J. (1989) The causes and consequences of surface water acidification. In: Morris R, Taylor EW Brown DJA, Brown JA (Eds) Acid Toxicity and Aquatic Animals. Soc Exp Biol Symp Seminar Series No 34, Cambridge University Press, pp 1-12.

McLusky, D.S. and Heard, V.E.J (1971) Some effects of salinity on the mysid *Praunus flexusosus* J mar biol Ass UK 51: 709-715.

McMahon, B.R. and Stuart, S.A. (1989) The physiological problems of crayfish in acid waters. In: Morris R, Taylor EW Brown DJA, Brown JA (Eds) Acid Toxicity and Aquatic Animals. Soc Exp Biol Symp Seminar Series No 34, Cambridge University Press, pp 171-199.

Morris, R,J., Lockwood, A.P.M., Dyball, D. and Bolt, S.R.L. (1987) Changes in the fatty acid composition of the gill phospholipds in *Gammarus duebeni* during moult: evidence for reduced permeability of the gill membrane. Comp Biochem Physiol 88B; 257-260.

Morritt, D. and Spicer, J.I. (1995) Changes in the pattern of osmoregulation in the brackishwater amphipod *Gammarus duebeni* Lilljeborg (Crustacea) during embryonic development. J Exp Zool 273: 217-281.

Morritt, D. and Spicer, J.I. (1996) The culture of eggs and embryos of amhipod crustaceans: implications for brood pouch physiology. J mar biol Ass UK 76: 361-376

Pace, F., Harris, R.R. and Jaccarini, V. (1976) The embryonic development of the Mediterranean freshwater crab *Potamon edulis* (=*P. fluviatilis*)(Crustacea, Decapoda; Potamonidae). J Zool Lond 180: 93-106.

Percy, J.A. (1985) Temperature tolerance, salinity tolerance, osmoregulation and water permeability of Arctic marine isopods of the *Mesidotea*(=*Saduria*) complex. Can J Zool 63: 28-36.

Potts, W.T.W. (1994) Kinetics of sodium uptake in freshwater animals: a comparison of ion exchange and proton pump hypotheses. Am J Physiol 266 (Regulatory integrative comparative physiology 35): R315-R320.

Potts, W.T.W. and Durning, C.T. (1980) Physiological evolution in the branchiopods. Comp Biochem Physiol 67B: 474-484.

Potts, W.T.W. and Fryer G (1979) The effects of pH and salt content on sodium balance in *Daphnia magna* and *Acantholeberis curvisrostris* (Crustacea: Cladocera). J Comp Physiol 129: 289-294.

Read, G.H.L. (1984) Intraspecific variation in the osmoregulatory capacity of larval, post-larval, juvenile and adult *Macrobrachium petersi*(Hilgendorf) Comp Biochem Physiol 78A: 501-506.

Shaw, J. (1960) The absorption of sodium ions by the crayfish *Astacus pallipes* Lereboullet III. The effect of other cations in the external solution. J Exp Biol 37: 548-556.

Shires, R., Lane, N.J., Inman, C.B.E. and Lockwood, A.P.M. (1994) Structural changes in the gill cells of *Gammarus duebeni* (Crustacea: Amphipoda) under osmotic stress: with notes on microtubules in association with the septate junctions. Tissue Cell 26: 767-778.

Spaargaren, D.H. (1989) adaptation to estuarine conditions in shore crabs *Carcinus maenas* (L.) in relation to body size. J Exp Mar Biol Ecol 129: 25 - 263.

Stobbart, R.H., Keating, J. and Earl, R. (1977) A study of sodium uptake by the water flea *Daphnia magna*. Comp Biochem Physiol 58A;: 299-309.

Sutcliffe, D.W. (1971) Sodium influx and loss in freshwater and brackish-water populations of the amphipod *Gammarus duebeni* Lilljeborg. J Exp Biol 54: 255-268.

Sutcliffe, D.W. (1975) Sodium uptake and loss in *Crangonyx pseudogracilis* (Amphipoda) and some other crustaceans. Comp. Biochem Physiol 52A: 255-258.

Sutcliffe, D.W. (1978) Water chemistry and osmoregulation in some arthropods especially Malacostraca. Freshwater Biological Association Annual Report 1978 .

Taylor, P.M. (1985a) The pattern of gill perfusion in two species of *Corophium* (Crustacea: Amphipoda) and its relation to environmental salinity. J Zool Lond (A) 205: 29-38.

Taylor, P.M. (1985b) Electrical potential differences and sodium ion fluxes across the integument of *Corophium volutator* (Crustacea: Amphipoda) a euryhaline hyperosmotic regulator. J Exp Biol 114: 471-491.

Taylor, P.M. and Harris, R.R. (1986) osmoregulation in *Corophium curvispinum* (Crustacea: Amphipoda), a recent coloniser of freshwater. I Sodium ion regulation. J Comp Physiol B 156: 323-329.

Thuet, P. (1988) Relation entre osmoregulation et activite d'ATPase Na^+-K^+ et d'anhydrase carbonique chez larves et post-larves de *Hommarus gammarus* (L) Crustacea: Decapoda. J Exp Mar Biol Ecol 115: 249-261.

Towle, D.W. (1990) Sodium transport systems in gills. In: Kinne RKH (Ed) Comparative aspects of sodium co-transport systems. Comparative Physiology Vol 7, Karger , pp 241-263.

Van den Brink, F.W.B., Van der Velde, G. and De Vaate, A.B. (1991) Amphipod invasion on the Rhine. Nature 352: 576.

Vangenechten, J.H.D., Witters, H and Vanderborght, O.L.J. (1989) Laboratory studies on invertebrate survival and physiology in acid waters. In: Morris R, Taylor EW Brown DJA, Brown JA (Eds) Acid Toxicity and Aquatic Animals. Soc Exp Biol Symp Seminar Series No 34, Cambridge University Press, pp

Weider, L.J. and Hebert, P.D.N. (1987) Ecological and physiological differentiation amongst low-arctic clones of *Daphnia pulex*. Ecology 68: 188-189.

Wood, C.M. Rogano MS (1986) Physiological responses to acid stress in crayfish (*Orconectes*): haemolymph ions, acid-base status and exchanges with the environment. Can J Fish Aquat Sci 43: 1017-1026.

Ambient Ions and the Voltage Across Crayfish Gills

L. B. Kirschner

Department of Zoology, Washington State University, Pullman, WA 99164 USA

2.1 Introduction

The potential difference across open-circuited amphibian skin (the TEP) is generated by Na^+ transport from medium to blood. Transport of other ambient ions contributes little although an effect of high [Cl^-] has been described (Harck and Larsen, 1986) and will be illustrated below. Ion transport systems in fishes and crustaceans resemble those in amphibians (Kirschner et al, 1973) but the electrical pictures are very different. The purpose of this review is to describe the contributions of Na^+, Ca^{2+} and Cl^- to electrogenesis in crayfish gills the procedures for measuring the TEP in crayfish have been recently described (Kirschner, 1994).

2.2 Experimental Results

2.2.1 Tap Water Medium and Amiloride

To provide background, figure 2.1 shows the TEP in an anesthetized frog immersed in tap water (TW; in mM, Na+ = 1.8, Ca^{2+} = 2.9, Cl^- = 1.9, pH 7.3). The TEP (V_{in} - V_{out}) was +48 mV, about the same as in isolated skins under simllar conditions (Ehrenfeld et al, 1985). When the medium was changed to TW containing TEA-Cl (20 mM) the TEP dropped to +35 mV, which is close to the Cl^- equilibrium potential (E_{Cl}). This reflects a large increase in Cl^- permeability due to opening of apical Cl^- channels (Harck and Larsen, 1986). The permeability change was reversible; on replacing the high-Cl^- medium with fresh TW the TEP returned to +46 mV, close to the original value. When amiloride (2 x 10^{-5} M) was added the TEP dropped rapidly to +14 mV, then more slowly to about -14 mV. The action of amiloride was reversible (not shown).

Thus, it is possible to reproduce, in the intact frog, the key elements of electrogenesis deduced from studies on isolated skins. It is interesting that there is no evidence that Ca^{2+} plays an important role, which is also in accord with earlier studies.

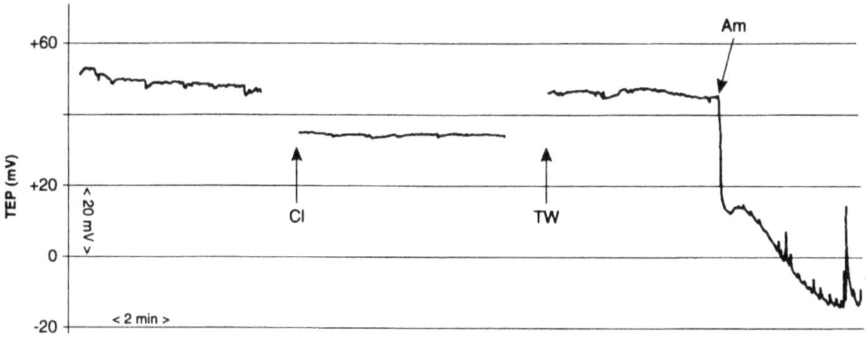

Fig. 2.1 The frog TEP in tap water. A single animal was initially in TW. At the first break in the recording the chamber was emptied and refilled with TW t TEA-Cl (20 mH).
At the second break it was emptied and refilled with TW. Amiloride was added without changing the solution.

A similar experiment produced very different results in crayfish (figure 2.2). The TEP in TW was close to O mV. When TEA-Cl was added it decreased, as in frog skin, but to -12 mV, which is far from ECl (about +60 mV). When the bathing medium was replaced with TEA-C1 (20 mM; Na^+ and Ca^{2+}-free)

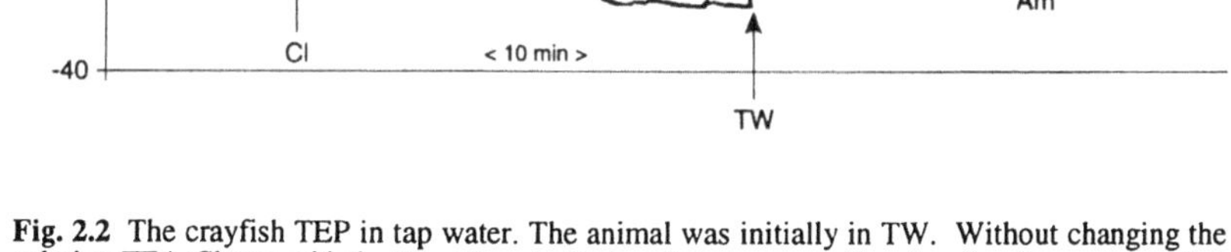

Fig. 2.2 The crayfish TEP in tap water. The animal was initially in TW. Without changing the solution TEA-Cl was added ~20 mH~. At the break in the recording the chamber was emptied and refilled with TEA-Cl (20 mM; Na^+ Ca^{2+}-free). This mediun was replaced by TW, and the TEP returned to its oriqinal value. Amiloride ~10^{-4} H) had no effect.the value dropped to -26 mV. Changing to TW brought the TEP to +4 mV, near the original value. Amiloride (10^{-4} M) had no effect, although this concentration blocks Na^+ influx under these conditions. ~Kirschner et al 1973 Ehrenfeld 1974).

2.2.2 The effect of Ca^{2+}

It has been shown that when crayfish are in an ion-poor medium (IP; in mM, KCl 0.1 $KHCO_3$ 0.1) the TEP is usually -20 to -40 mV. This is probably a diffusion potential set up by leakage of NaCl from hemolymph to medium (Kirschner 1994). Stepwise addition of Ca^{2+} resulted in depolarization, and the concentration-dependence is shown in figure 2.3. The TEP approached an upper limiting value around 0 mV at $[Ca^{2+}]$ = 1 - 2 mM, close to the TW value.

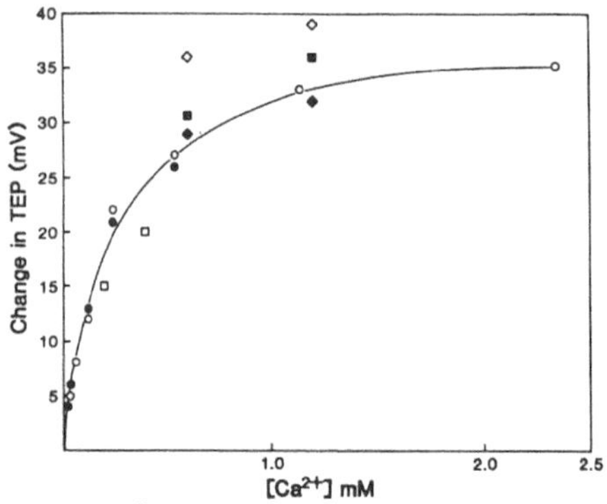

Fig. 2.3 Ca^{2+}-dependence of the crayfish TEP. Each type of symbol represents values from a single animal. The line was fitted by eye through the data (circles) from two animals exposed to increasing concentrations over the range 0.02-2.4 mM (from Kirschner, 1994).
Threshold for the effect was 10-20 mM, and the concentration inducing half-maximal change was 0.1 - 0.2 mM. The values are similar to those found in *Austropotamobius pallipes* (Greenaway, 1972).

2.2.3 The Effect Of Na^+

Figure 2.4 shows results from a single animal in IP solution; the initial TEP was -19 mV. Incremental additions of Na^+ (sulfate) produced stepwise depolarization with a final value of +70 mV at 78 mM. A series of 7 animals gave qualitatively similar results.

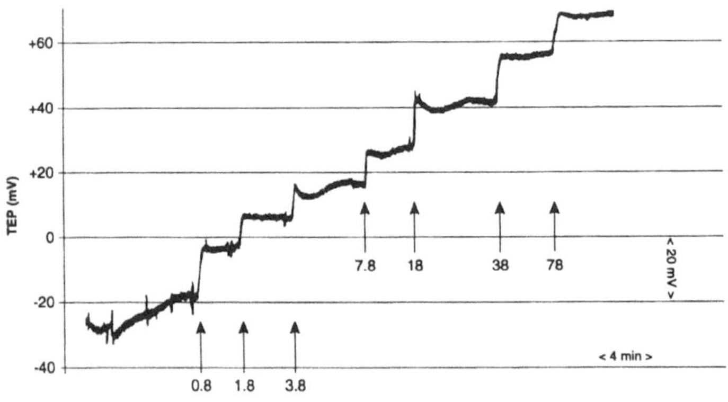

Fig. 2.4 Na^+-dependence of the crayfish TEP. The animal was initially in IP medium with a TEP of about -19 mV. At the first arrow Na_2SO_4 was added to the medium (final $[Na^+]$ was 0.8 mM). At each subsequent arrow the concentration was increased. The $[Na^+]$ after each increment is shown below the corresponding arrow.

Since the TEP has a substantial diffusional component these data must be corrected for changes in the Na^+ gradient. When the correction was applied to the data in figure 2.4 there remained a non diffusional TEP change approaching +25-+30 mV as an upper limit, with half maximal change occurring at about $[Na^+]$ = 2 mM.

In the absence of Ca^{2+}, amiloride had a substantial effect on the Na^+-induced depolarization at low ambient $[Na^+]$. In the range 0 - 4 mM amiloride (10^{-4} M) reduced the extrusion from 17 mV to 7 mV.

2.2.4 The Effect of Chloride

Two experiments suggest that Cl^- is also electrogenic, but with an effect opposite that produced by the two cations. As shown in figure 2.2 addition of 20 mM Cl^- shifted the TW TEP in the negative direction. The observation was extended in a single animal in which stepwise increments in $[Cl^-]$ produced incremental hyperpolarization. The second experiment demonstrated that the change in TEP with increasing $[Na^+]$ was smaller when the anion was Cl^- than when it was SO_4^{2-} as shown in figure 2.5. The values shown have not been corrected for changes in diffusional components (which differ slightly for the 2 anions), but the reduction in the presence of Cl^- was substantial even after the correction was made.

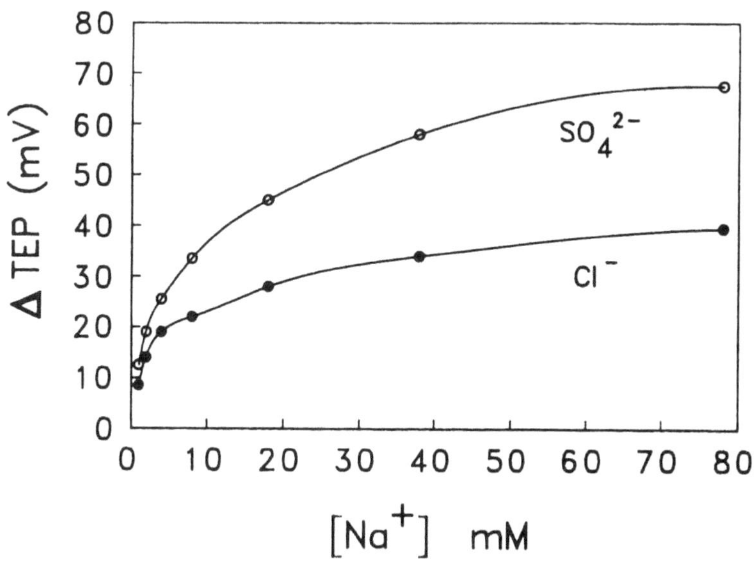

Fig.2.5. The effect of Cl^- on the Na^+-dependent TEP. The protocol was as in fig. 2.4, but the dependent variable here is the chance in TEP. Each point is the average of 2 experiments (different animals) with the order of addition reversed; i.e. the Na_2SO_4 series first in one experiment, NaCl series first in the other. The changes are clearly smaller in NaCl.

2.3 Discussion

2.3.1 The TEP and Ambient Ions

There are clearly at least four variables that contribute to the TEP in crayfish. One of these involves the diffusive movement of NaCl from hemolymph to medium across gills with a $P_{Cl}/P_{Na} < 1$, usually 0.25 - 0.4. In the absence of transportable ions, i.e. in an IP solution, this results in a TEP of -20 to -40 mV (Kirschner, 1994). The second factor is ambient [Ca^{2+}] which generates a non linear, concentration-dependent change that makes the hemolymph less negative. The threshold concentration is about 10 μM, that for halfmaximal change 100 - 200 μM, and the TEP approaches an upper limiting value around 0 mV at 1 - 2 mM. Sodium in the medium has the same qualitative effect as Ca^{2+}, but the parameters are different. The threshold is >100 μM and the half-maximal change occurs at [Na^+] = 2 - 3 mM. Finally, ambient Cl^- causes the opposite effect. The TEP becomes more negative as [Cl^-] increases; effective concentrations appear to be in the same range as for Na+ (unpublished data).

Most natural FWs contain all three ions, but the sensitivity of the TEP to Ca^{2+} is an order of magnitude greater than to Na^+ or Cl^- (i.e. the effective [Ca^{2+}] is much lower). This suggests that in most FW experiments the two factors that determine the TEP are ambient [Ca^{2+}] and the relative permeability (P_{Cl}/P_{Na}) of the gill. Both are likely to differ among laboratories and groups of animals, but the combination appears to account for the fact that most values reported are in the range 0 ± 10 mV. The very different electrical picture seen in amphibian skin is likely due to the fact that Ca^{2+} has little or no effect on the skin's electrical behavior.

2.3.2 Implications for Ion Transport Systems

Calcium uptake in crayfish was shown to require active transport (Greenaway, 1972), and the electrical picture suggests that transport is electrogenic. One possible mechanism is that transfer across the apical membrane is via a channel with basolateral transfer mediated by a Ca^{2+}ATPase (Kirschner,1994) as had earlier been proposed for fish (Perry and Flik, 1988). But there has also been described a Ca^{2+}-H^+ exchanger in crustaceans (Ahearn and Franco, 1990), and such a system can also account for the observations, since it carries net charge into the cell.

The effect of ambient Na^+ on the TEP indicates that inward transport of this ion is also electrogenic. This immediately rules out the operation of a 1:1 Na^+-H^+ exchanger in the apical membrane. The remaining alternatives are the same as for Ca^{2+} an apical Na^+ channel, (and proton pump) as in amphibian skin or an asymmetric exchanger. A $2Na^+$-$1H^+$ exchanger has been described in many crustaceans (Ahearn et al, 1990). These studies were made on internal organs of marine animals, but one was on the gills of the crab *Carcinus maenas* which, although basically marine, can live in dilute media (Shetlar and Towle, 1989). Amiloride, which can often distinguish between exchanger and channel, is of no help in this case. Its K_i in aquatic animals (in the order of 10^{-5} M) is much higher than for most channels but lower than for most exchangers. However, other factors argue against the operation of the exchanger in crayfish. First, in salt-deleted animals, and with Cl^- absent, the apical exchange has been shown to be 1:1 (Kirschner et al, 1973, Ehrenfeld, 1974), not 2:1. In addition, Na^+ movement via the crustacean exchanger is blocked by Ca^{2+} at 0.5 mM (Ahearn et al, 1990), but $[Ca^{2+}]$ up to 5 mM has been shown to have little effect on Na^+ influx (Shaw, 1960). Finally, for an animal in a medium with $[Na^+] = 0.9$ mM and pH 7, $[Na^+]$ cell would be < 2.7 mM (for the exchanger to continue to move Na^+ inward). This is as low as has ever been reported for a similar epithelium (Harvey and Kernan, 1984; for frog skin with $[Na^+] = 0.1$ mM in the medium) and would require that the basolateral Na^+-K^+ pump operate at >90% efficiency. It should be emphasized that crayfish can transport Na^+ from media with $[Na^+]$ considerably below 1 mM and pH lower than 7 where $[Na^+]$ cell would be even lower. Overall, the evidence seems to favor the operation of the channel-pump model as in frog skin.

The limited information on the effect of Cl^- on the TEP suggests that an electrogenic mechanism might be responsible for transport in these experiments. Such a system has also been shown to operate in frog skin (Larsen et al, 1992).

References

Ahearn, G.A., and P. Franco (1990). Sodium and calcium share the electrogenic $2Na^+$-$1H^+$ antiporter in crustacean antennal glands. Am. J. Physiol. 259:F758-F767.

Ahearn, G.A., P. Franco, and L.P. Clay (1990). Electrogenic 2 Na^+/1 H^+ exchange in crustaceans. J. Memb. Biol. 116:215226.

Ehrenfeld, J. (1974). Aspects of ionic transport mechanisms in crayfish *Astacus leptodactylus*. J. Exp. Biol. 61:57-70.

Ehrenfeld, J., F. Garcia-Romeu, and B.J. Harvey (1985). Electrogenic active proton pump in *Rana esculenta* skin and its role in sodium ion transport. J. Physiol. 359:331-355.

Greenaway, P. (1972). Calcium regulation in the freshwater crayfish *Austropotamobius pallipes* (Lereboullet). J. Exp. Biol. 57:471-487.

Harck, A.F., and E.H. Larsen (1986). Concentration dependence of halide fluxes and selectivity of the anion pathway in toad skin. Acta. Physiol. Scand. 128:289-304.

Harvey, B.J., and R.P. Kernan (1984). Intracellular ion activities in frog skin in relation to external sodium and effects of amiloride and/or ouabain. J. Physiol. 349:501-517.

Kirschner, L.B. (1994). Electrogenic action of calcium on crayfish gills. J. Comp. Physiol. 164B:215-221.

Kirschner, L.B., L. Greenwald, and T.H. Kerstetter (1973). Effect of amiloride on sodium transport across body surfaces of freshwater animals. Am. J. Physiol. 224:832-837.

Larsen, E., N. Willumsen, and B. Christoffersen (1992). Role of proton pump of mitochondria-rich cells for active transport of chloride ions in toad skin epithelium. J. Physiol. 450:203-216.

Perry, S., and G. Flik (1988). Characterization of branchial transepithelial calcium fluxes in freshwater trout, *Salmo gairdneri.* Am. J. Physiol. 254:R491-R498.

Shaw, J. (1960). The absorption of sodium ions by the crayfish *Astacus pallipes* lereboullet III. The effect of other cations in the external solution. J. Exp. Biol. 37:548-556.

Shetlar, R.E., and D.W. Towle. (1989). Electrogenic sodiumproton exchange in membrane vesicles from crab (*Callinectes sapidus*) gill. Am. J. Physiol. 257:R924-R931.

Regulating the Micro-environment of Ion Transporting Epithelia: A Comparative Approach

K.L. Shephard*

The Clore Laboratory for Life Sciences
The University of Buckingham
Buckingham, Bucks MK18 1EG
UK

* present address
LSU, Southampton. SO17 1BG, UK

3.1 Introduction

Epithelia, epidermal or otherwise, that transport significant quantities of ions share a range of common characteristics. They include the presence of polarised surface cells, often with numerous mitochondria, a range of ion pumps and channels and a complex surface architecture. Almost universally research into functional morphology of epithelia concentrates on the cells themselves and rarely delves into the external space that borders the cell's apical surface. This space is the surface's micro-environment and the physical and chemical conditions that prevail within it can be described as its micro-climate.

It is often assumed that epithelial ion transport should be enhanced if the micro-climate of its micro-environment is as uninfluenced by the activities of the epidermis as possible. The multifunctional fish gill, for example, is well perfused, well ventilated and exposes a very large surface area to incoming water. A micro-environment that is anything other than clean water would seem at first analysis to be inappropriate. To anything other than casual observation, however, it is clear that the gill has a great potential for regulating its micro-environment. The epidermis of gill bars and filaments, for example, often contain large numbers of goblet cells, capable of discharging a considerable volume of mucus; capable in turn of swelling to occupy the bulk of the micro-environment of the gill surface and imposing a micro-climate considerably different from that obtained by the passage of respiratory water over the gill surface.

Another interesting example is offered by the mammalian trachea. The 'airway' is of necessity exposed to air with a range of temperature and humidity; in some circumstances to dry cold air able to evaporate a considerable quantity of water (Barbet *et al* 1988) prior to saturation. A dry impermeable airway surface would seem a desirable attribute to any land animal with water conservation problems. Yet the maintenance of a dry, impermeable airway surface is not an option used by any terrestrial chordate. It seems likely that the requirement that the airway remains clean and unblocked dictates the need for mucociliary transport, which in turn requires the surface to be wet. If cilia are dried, even momentarily, they do not survive. Water must be present and it is likely that ion transport systems are used by the airway for this purpose. Water must also be provided to hydrate the mucus that is required for mucociliary transport, and to modify the inspired air that must reach the alveolae fully humidified at body core temperature if it is to avoid dehydrating this delicate surface. Indeed the mammalian airway is the site of considerable heat and water exchanges of great importance to all mammals and particularly so for those that live without access to free water and who are so ably adapted to water conservation. This degree of micro-environment management must require precise regulation.

It is clear, therefore, that at least some ion transporting epithelia put a great deal of effort into regulating the micro-environment of their surfaces. Indeed, for some it is probably true that a major component of ion transport is involved in this regulation. This review looks at some of the processes used by animals to regulate the micro-environment of ion transporting epithelia. It will concentrate on three aspects: regulation of the movement of water across the epithelia; regulation of secretion by goblet cells; and regulation of the swelling and dispersal of secretions, in particular those of goblet cells.

3.2 Water movements across epithelia

The mammalian airway provides an excellent epithelium for the analysis of water-transport regulatory processes. Research to understand the genetic disorder cystic fibrosis, for example, ensures that an abundance of information is available. Cystic fibrosis is, in particular, a disease characterised by inadequate hydration of the airway surface.

Water is thought to cross the airway surface primarily by following osmotic gradients established by ion pumping. Both active chloride secretion and active sodium absorption are well characterised in a range of mammalian airways (Olver *et al* 1975, Frizzel 1988). Stimulating active chloride secretion in the canine airway does increase fluid secretion (Welsh *et al* 1980) while stimulating active sodium absorption does increase fluid absorption (Nathanson *et al* 1983). Much remains to be discovered about the mechanisms, however, and the location or even the presence of local osmotic gradients remains to be demonstrated. Nevertheless ion-transport generated osmotic gradients remain at the centre of our current understanding of water transport regulation in the mammalian airway. The upper airway of mammals is primarily an absorptive surface. Reduction in the cross-sectional perimeters of airways between lung and nose coupled with upwards mucociliary transport mean that fluid is in excess in this region (Kilburn 1968). It is likely that the lower, smaller airways are primarily secretory (Al-Bazzaz 1994). These processes are probably regulated by a variety of physiological systems. Certainly a range of pharmacological agonists stimulate active transport (e.g. bradykinin, isoprotereol, ATP, Widdicombe & Widdicombe 1995) but the processes of regulation are not known. What characteristics of the layer of airway surface liquid are monitored in order to regulate the ion transport that will produce the osmotic gradients that will drive water movement?

We should not forget the ability of passive control mechanisms to regulate fluid at the airway surface. After all a significant contribution to fluid balance in biological systems is made by simple equilibria between hydrostatic pressure and colloidal osmotic pressure. This equilibrium, for example, largely dictates the scope of fluid flow at the capillary level. Perhaps similar equilibria exist at the airway surface?

Evaporation of water at the airway surface may provide a powerful and, to some extent, self-regulating, modulator of osmotic pressure in airway surface liquid. Breathing in dry cold air evaporates water from all surfaces until the air is in heat and moisture equilibrium with the body. Although this activity is concentrated in the upper airway (Cole 1982), during exercise and in cold dry conditions it can progress deeper (Daviskas *et al* 1991, McFadden 1983, McFadden *et al* 1985). It certainly leads to an increase in the osmotic pressure of airway surface liquid (Man *et al* 1978) in anaesthetised canine trachea, and to increase in ion concentrations (Shepard & Rahmoune 1994) in isolated guinea pig trachea. In the isolated guinea-pig trachea evaporation-induced changes in airway surface liquied lead to an incease in the movement of water from the tissue into the surface layer to support this evaporation. This indicates that the tissue is permeable enough for evaporation to drive water into the airway surface liquid (Shephard & Rahmoune 1994). Reversing the osmotic pressure gradient drives water in the opposite direction (Rahmoune & Shephard 1994, 1995). Evaporation from the trachea is likely to cool its surface. Expiration of warm humid air over this cold surface will certainly capture much of the evaporated water and may reduce the potential for evaporation to increase the osmotic pressure of airway surface liquid, but most animals lose some moisture during breathing.

An additional passive contribution to the maintenance of a moist airway surface may come from the hydrostatic pressure of capillaries within the airway (Raphael *et al* 1989) particularly during inflammation (Widdicombe & Widdicombe 1995) but the situation is by no means clear. Likewise the potential involvement of surface tension effects, originating at the huge surface offered by cilia, in maintaining a layer of water on the airway surface in opposition to adverse osmotic pressure gradients is unknown. Certainly the surface tension is real and should balance a considerable osmotic pressure (the matric contribution to water potential used by plant physiologists) but the system is not readily accessible to experimentation and remains largely theoretical. A very thick layer of water over semi-dry cilia, for example, would be very vulnerable to drying out altogether.

3.3 Secretion by goblet cells

Epidermal tissues also secrete mucus to modify their micro-environment. This review will look towards this activity in the fish gill: a tissue which demonstrates almost no mucus on its surface under some circumstances (Handy & Eddy 1991) but a great amount when stressed by exposure to heavy metals in solution (McDonald & Wood 1993, Westfall 1945), by acid rain and aluminium (Muniz & Leivestad 1980), by silt-laden water (Roberts & Bullock 1980) and by a large number of other factors. The fish gill should provide an excellent preparation to study the processes used by animals to regulate secretion of mucus. It will be necessary, however, to refer to other tissues where our understanding of some matters has advanced further. An advance driven not by the convenience of the preparation for study, but rather by the immediate human consequences of malfunction, as in mammalian models of the asthmatic human airway. Not only does the fish gill provide an excellent comparative research tool, but there are also questions relating to secretion of mucus by this tissue important in their own right. Stressed fish produce mucus which limits the uptake of therapeutic and prophylactic agents required for modern fish husbandry (O'Grady 1991). An understanding of the regulatory mechanisms in this system is a necessary and useful goal.

This review will also focus on one cell type in fish gills: goblet cells. Other cell types may contribute to mucus. In particular malpighian cells are known to secrete a distinctive glycocalyx (Hughes & Wright 1970, Whitear 1970) but there is very little information about how this process is regulated, in fish gill or any other tissue. The goblet cells in fish gill have been extensively described both morphologically and histochemically (Laurent 1984) and have much in common with mammalian goblet cells seen in airway and intestinal tissue. It is possible that there are several types of goblet cell in the fish gill. Certainly subpopulations of goblet cells with differing histochemical and morphological characteristics exist in fish skin (Whitear 1986, Mittal *et al* 1995) and the presence of carboxylated and sulphated glycoproteins in mucus granules varies from cell to cell in skin goblet cells (Blackstock & Pickering 1982).

There is some information about the mechanisms whereby goblet cells release their mucus in fish. Goblet cells store condensed glycoprotein in membrane-bound granules which normally burst during the process of exocytosis, but which may on occasions be carried, unburst, into released mucus (Fletcher & White 1973). Goblet cells in fish gills are likely to be merocrine secretors as are mammalian goblet cells (Pickering 1976) (i.e. they release mucus granules at their apical surfaces while manufacturing more) but the process has been little studied. More information is known about goblet cell degranulation from work on isolated mammalian tracheal-epithelium (Davis *et al* 1992). Individual degranulations were observed in this preparation using high resolution Normaski light microscopy. Degranulations were resolved into non-stimulated base-line secretory events (occurring at a rate of 0.05 secretory events s^{-1}) and more rapid but variable stimulated secretion. The technique promises to yield much more information on this important process and should be applicable to a range of tissues and animal types.

Control or regulation of mucus release falls naturally into two categories. Control of the mechanisms through which degranulation processes occur and regulation of goblet cell proliferation and turnover. In reality the distinction may not be clear cut as several factors that influence proliferation of cells also influence degranulation processes.

Stressful circumstances are known, on occasions, to elicit a rapid and extensive release of mucus from the fish gill. Capture and handling (Shephard 1992a, b), suspended solids in water passing over the gills (Stuart 1953); irritants such as acidic water (McCahon *et al* 1987) all elicit rapid mucus release by the gill. The mechanisms involved, however, are far from clear. Mucus release by the gill may be a neurally mediated reflex. The epidermis of the fish gill filament is known to be extensively innervated (reviewed by Nilsson 1984) but there is no confirmation of this process in the gill. Also likely is the direct involvement of catecholamines released from chromaffin tissue. Such release is known to occur in response to a wide range of physical and chemical disturbances including capture, handling and hypoxia (Butler *et al* 1978). Again there is no direct confirmation that this process regulates mucus release in fish gills. For such direct information we must turn to goblet cells in mammalian systems. Considerable advances in our knowledge of the regulation of secretion of mucus are occurring as a result of the use of a human colonic goblet-cell line as monolayers of differentiated cells (Cl-16E). Mucin secretion by these cells follows transient increases in intracellular Ca^{2+} produced via Ca^{2+} influx from the external medium (indicating the involvement of receptor linked Ca^{2+} channels) and from intracellular stores. Links to the cyclic AMP pathway are also likely (Bouhanna *et al* 1994). The same preparation provides information about the receptors present on goblet cell membranes. It has long been known that goblet cells in the epidermis of the mammalian airway primarily respond to cholinergic stimulation but maintain some sensitivity to adrenergic activity. A small secretory response remains after adrenoceptor and cholinoceptor blockage (Peatfield & Richardson 1983). This non-adrenergic, non-cholinergic (NANC) neural mechanism includes an orthodromic (efferent) pathway in which neuropeptides (including neuropeptide tyrosine, NPY; and vasoactive intestinal peptide, VIP) are co-localised and co-released with acetylcholine (Rogers 1995). Cultured human colonic goblet cells correspondingly respond to carbachol (a cholinergic agonist) by elevating intracellular Ca^{2+} (blocked by atropine - indicating the muscarinic nature of the cholinergic receptor) and secreting mucins. The secretory response to carbachol is potentiated by VIP which stimulates the cAMP pathway (Bouhanna *et al* 1994). This combination of multiple receptors and multiple intracellular signalling pathways clearly involved in the regulation of airway goblet cells, is also seen in mammalian pancreatic ducts (Hootman & Deondarza 1995) and in mucous glandular cells in laryngeal gland (Tanaka *et al* 1995). It appears to be a common feature where these activities have been investigated.

The NANC neural mechanism in airway secretion also includes a sensory-efferent (or local-effector) pathway, sensitive to capsaicin, the active ingredient of peppers (Maggi & Meli 1987). Stimulation of sensory receptors in airways causes the release of neuropeptide neurotransmitters including the tachykinins substance P and neurokinin A at efferent collateral terminals resulting in secretion of mucus, particularly by submucosal glands (Ramnarine *et al* 1994) but also by goblet cells (Rogers 1994). The presence of this sensory-efferent pathway to mucus secretion is particularly relevant to studies of chronic inflammatory bronchial diseases (such as asthma and bronchitis, characterised by hypersecretion of mucus) where there is some evidence for increased sensory-efferent neural activity (Rogers 1995, Hayes *et al* 1995).

Receptor-mediated secretory activity of goblet cells is also induced by phosphorylated nucleotides ATP, ADP and AMP via the P-2 type receptor (Merlin *et al* 1994, Davis *et al* 1992). Activation of these receptors results in fusion of granule membranes, mucin release and opening of chloride channels. These events are not associated with cell damage and represent normal secretory activity. ATP is known to be a neurotransmitter co-secreted by adrenergic and cholinergic neurones (Burnstock 1986) and may be a neurotransmitter released by neurones in the non-adrenergic, non-cholinergic system in mammals. This work is related to the mechanisms that enable secreted mucin to swell and disperse and will be considered in more depth below.

The regulation of goblet cell proliferation and turnover, in contrast to the regulation of degranulation processes, has been extensively examined in fish (Shephard 1994) but primarily in fish skin. Several hormones are known to have pronounced influences on the dynamics of goblet cells in fish skin and the factors that influence hormone levels have been the subject of a great deal of research. The hormone prolactin, for example, long associated with epithelial ion regulation (Nicholl & Bern 1971), is known to have dramatic influences on epidermal goblet cells. Administration of prolactin induces a thickening of the epidermis and proliferation of goblet cells in many species (Ogawa 1970, Mattheij & Stroband 1971, Olivereau & Lemoine 1971, Marshall 1976, Wendelaar Bonga and Meis 1981). Hypophysectomy reduces the number of goblet cells in fish skin and this can be reversed by injecting prolactin (Marshall 1976). Prolactin levels *in vivo* are influenced by a range of environmental variables including calcium concentrations and osmolarity (Wendelaar Bonga & Meis 1981). Other hormones are known to influence dynamics: catecholamines (Ojha & Munshi 1974); oestrogens and androgens (Pickering & Richards 1980); testosterones (De Ruiter & Mein 1982, Burton & Everard 1991) and cortisol (Marshall 1979).

The effects of cortisol and stress are particularly well understood due to the research of Wendelaar Bonga and his associates in Nijmegen. A single dose of cortisol administered in the food of rainbow trout produces elevated blood-cortisol levels for 24 hours. Among many changes in skin structure were increased apoptosis of goblet cells and the presence of newly differentiated mucous cells near to the skin surface (Iger, Balm, Jenner & Wendelaar-Bonga 1995). These changes closely parallel many of those induced by exposing the fish to stressful conditions. Exposing the same species to acidified water (pH 5.0), for example, resulted in increased apoptosis (and necrosis), mucous cell elongation and the appearance of newly differentiated mucous cells near to the skin surface in conjunction with elevated blood-cortisol levels (Iger, Balm, Jenner & Wendelaar-Bonga 1994). Blood-cortisol levels are also increased in fish stressed by disease and experimental administration of bacterial lipopolysaccharide to tilapia (*Oreochromis mossambicus*) *in vivo* induced the same combination of elevated blood cortisol and proliferation of mucous cells in the skin (Balm, Vanlieshout, Lokate & Wendelaar-Bonga 1995). Goblet cell proliferation is even increased by routine handling of fish (Pickering & Macey 1977).

Changes in salinity may have a more subtle effect on goblet cell dynamics. In general the abundance of goblet cells on gills and on skin increases as salinity decreases (Burden 1956; Vibabhadrachari 1961; Ahuja 1970; Wendelaar-Bonga 1978; Laurent 1984). The combined effects of prolactin and cortisol are probably involved. It has also been suggested that metaplastic changes contribute to the change in abundance of goblet cells; goblet cells have been suggested to transform into chloride cells when ambient salinity is raised (Zaccone *et al* 1981). Metaplastic changes, involving goblet cells, are known to occur in the mammalian airway where goblet cells do differentiate into ciliated epithelial cells in normal conditions but develop into mature secreting cells in response to chronic exposure to airway irritants (Rogers 1994). Chronic exposure to environmental stressors is known to produce acclimatory responses by fish which involve goblet cells on the gill. Recently reviewed by McDonald & Wood (1993) these responses include hypertrophy of cells in response to aluminium (McDonald *et al* 1991).

Areas of physiological functioning that fall between established disciplines generally receive less attention than they deserve. Release of mucus is clearly an important stress response by many animal tissues and research has emphasised neuronal and hormonal regulatory processes. Interactions between these processes and those of the immune system are only slowly being uncovered by immunologists more involved in specific and non-specific responses to disease. In fish gills, as in other animal systems, the area of interaction relates primarily to the inflammatory processes that have such profound influences on tissue functioning and which develop in response to the tissue damage that is often associated with physically or chemically mediated stress. Mucus release is, for example, a characteristic response of the excessive, and inappropriate, inflammation of the human airway associated with asthma. Triggers, which include allergens or stresses such as cold air and air pollutants, induce the release of a range of mediators (including histamine, the protein derived cytokines and the lipid derived leukotrienes) from many cells types. These inflammatory mediators may have direct effects on goblet cells (for example, leukotrienes, Hoffstein *et al* 1990; prostaglandins, Adler *et al* 1990) but they also attract both eosinophilic and neutrophilic granulocytes resulting in goblet cell degranulation due to the activity of proteinases (Klinger *et al* 1984), reactive oxygen species and other cell products (Adler *et al* 1990). There are parallels in the fish gill. Diapedesis of several cell types is known in the gills (and gut) of stressed fish. Cells moving into the epidermis include eosinophils (Powell et al 1990; Barnett et al 1996), macrophages and lymphocytes (Iger, Balm & Wendelaar Bonga 1994). Recent research by Barnett *et al* (1996) demonstrates that eosinophilic granulocytes accumulate in the filament epidermis in *Oreochromis mossambicus* in response to mild and routine handling stress and are also attracted to sites of micro-injection of histamine. Their potential role in mucus release is still to be confirmed (Figure 3.1). It is interesting to note, however, that injection of murine interleukin 1 into *Oreochromis mossambicus* causes a proliferation of epidermal mucous cells (Balm *et al* 1995) and that carp macrophages and neutrophilic granulocytes do produce an interleukin 1-like molecular with T-cell proliferating potency which has structural similarities to mammalian interleukin 1 (Verburg van Kemedade *et al* 1995).

3.4 Exocytosis, swelling and dispersal of mucus

In the last section the factors that trigger and regulate the release of mucus by goblet cells were considered. Here we consider how ion transporting epithelia regulate the subsequent processes. Exocytosis of mucus (the process whereby the contents of intracellular granules or vesicles become externalised) is, however, inextricably linked to subsequent swelling and dispersal, as the processes that dictate all 3 have much in common. This arises as a consequence of the molecular nature of mucins. Mucins are huge, linear, polyanionic glycoproteins with molecular weights as great as 10^7 and molecular lengths of several mm (Carlstedt & Sheehan 1989). When released from goblet cells these molecules expand (potentially indefinitely) and entangle with one another to occupy space. The degree of expansion by some mucous granules is as great as 600 fold prior to dispersal (Verdugo 1990, Kapeleta *et al* 1996) and this marks the considerable difference that exists between exocytosis of mucus and that of other cell granules (heparin released by mast cells, for example, is thought to expand 2-4 fold prior to dispersal; Fernandez *et al* 1991). The processes that drive exocytosis in this granular system almost certainly drive swelling of the resultant mucous gel, annealing of adjacent mucous gels and, ultimately, dispersal. A key property of mucins is their charged, polyanionic nature at near-neutral pH. Negative charges are neutralised by associated hydrogen ions at low pH but, almost certainly by divalent calcium ions at near neutral pH, *in vivo*. The presence of multivalent ions to neutralise these charges seems to be a necessary corollary of the highly condensed state of these macromolecules within intracellular granules. Divalent ions shield potentially repellant anionic sites from adjacent molecules and effectively stabilise the condensed state. Removal of divalent ions would be expected to result in mutual repulsion of adjacent charged sites and expansion of the granules contents. Replacing divalent ions by monovalent ions would satisfy a charge requirement, but would result in a considerable increase in the internal osmotic pressure of this 'Donnan' system. The fundamental nature of these interactions was described in principle by Donnan and associates at the turn of the century and by physical chemists for polyelectrolyte ion-exchangers in mid-century (Hellferich 1962) The application of ion-exchange theory to biological processes has taken longer to gain acceptance. In principle the process that drives exocytosis, swelling and dispersal of goblet cell mucus is inherently incorporated into a charge-shielded, condensed polyanionic polymer network, constructed within a membrane-bound vesicle within the goblet cell and effectively storing the energy required for expansion. This is the Jack-in-the-Box model of mucus exocytosis (Verdugo 1991). Jack is sprung from his box by removing calcium and replacing it with sodium. Support for the general model also comes from an analysis of other secretory systems. Several types of secretory granule contain polyionic polymers shielded by divalent ions: for example, polyanionic heparin is present within mast cell granules in conjunction with cationic, divalent histamine (Uvnas & Aborg 1977).

For goblet cells, however, and indeed for other secretory cells, details of the process whereby calcium is removed or replaced by monovalent sodium are sketchy. The precise development role of the goblet cell secretory-pore, through which mucus expansion must pass, is controversial. It has been proposed that the formation of a high conductance secretory pore which links the mucous granule to the external world is necessary to allow calcium ions to be replaced by sodium ions. This logical deduction is supported by extensive work on exocytosis of cortical granules from the sea urchin egg. Patch-clamp techniques, for example, suggest that pore formation precedes the swelling of the granule (Zimmerberg *et al* 1987, Breckenridge *et al* 1987). Nevertheless there is also evidence that the membranes surrounding secretory granules are under tension prior to 'docking' with the apical membrane and pore formation (Monck et al 1991) at least in the mast cell. This implies that some process of water accumulation or gel expansion within the granule occurs prior to pore formation.

A clue to this process may be provided by the mucous granules secreted by terrestrial slugs (Figure 3.2). These granules are released in large quantities by the slug and, under certain conditions, burst in the mucus surrounding the animal (in particular, between the animals foot and the substrate), rather than during the exocytotic process. As the granules do not dock with the goblet cells apical membrane other processes are presumably responsible for the initiation of the expansion process. Some of the processes involved have been elucidated by Deyrup-Olsen and colleagues (Deyrup-Olsen *et al* 1992) and depend on the observations that mucous granules are enclosed within 2 separate membranes, each with distinct properties; that granule bursting and swelling is specifically triggered by ATP; and that agents that specifically inhibit K^+Cl^- cotransport or block calcium channels prevent bursting. These authors propose that ATP stimulates K^+Cl^- cotransport into the granules and that the resultant swelling or membrane tension triggers the opening of calcium channels. Loss of calcium then stimulates rapid granule expansion and bursting. Whether or not this sequence of events offers an advantage to the slug is by no means clear. It has been suggested (Kapeleta *et al* 1996) that by swelling granules outside the body wall, rather than within it, much of the water for expansion probably comes from the animal's environment, and not from its body; an important consideration for a terrestrial animal with a living and highly permeable epidermis.

The process of swelling is, therefore, driven using energy incorporated into the condensed molecules and expressed by charge repulsion by adjacent molecules. The expanding gel is inevitably a rapidly changing Donnan system, with a dynamic equilibrium established between the electrical and chemical potentials of diffusible ions and an osmotic pressure difference between the gel and its surroundings. Equilibria of this sort have been described for a variety of chemical and biological systems (Helfferich 1962, Shephard 1987a, 1989) and in particular the effects of monovalent and divalent ions on the nature of the equilibrium are noteworthy. Swelling pressures (the difference between the internal and external osmotic pressure) for example, are much greater when the polyion has equilibrated with monovalent ions than with divalent ions. More monovalent ions are required for charge equilibrium than divalent ions and the osmotic pressure difference is, consequently, larger. It has been suggested, however, that while the direction of water movement into the system can be described in simple ion-exchange terms its kinetics cannot. Swelling is known to occur very rapidly, even explosively, in some granules (Deyrup-Olsen 1983) and diffusion of water along an osmotic pressure gradient may not account for the rate of swelling. Charge interactions have been invoked to explain the discrepancy (Verdugo 1991) but further evidence is needed to clarify the situation.

The process of mucus swelling is thought to be a biological example of a phenomena studied by polymer scientists known as 'phase transition' (Tanaka 1981). Polymer gels exist in both condensed and expanded (hydrated) phases. Both are relatively stable and reversible transition between them occurs as a change in the molecular configuration of the polymer network. Membranes surrounding the gel may dictate the phase that the gel exists in but only by controlling the movement of ions on which phase stability depends. Membranes do not physically restrain an expandable gel. Without a membrane, the expanded mucin matrix of mammalian respiratory goblet cells can be recondensed and stabilised by raising the external calcium concentration to 100mM at pH4 (Verdugo 1991). Similar results have been obtained with slug mucous granules (Verdugo 1991) and mast cell granules (Fernandez *et al* 1991).

An epidermis which releases expanding mucins, or intact mucous granules to burst later, must either provide a suitable micro-environment to support swelling of the mucous gel, or rely on the bulk environment to do so. In general terms the micro-environment must provide water and sufficient monovalent ions to replace calcium within the gel, but the extent and rate of swelling is then encouraged by an osmotically dilute medium (Helfferich 1962) and discouraged by an abundance of divalent or trivalent ions (Shephard 1987a). The effects of a lack of water are obvious when a dehydrated slug attempts to walk on a dry surface: it is rapidly immobilised in sticky mucus. Slug skin in fact contains specialised channel cells through which body fluid passes to facilitate expansion of mucus; ensuring that hydrated slugs can move (Martin & Deyrup-Olsen 1986). Fish presumably have no problem in providing water and this subject has been described for mammalian airways in section 3.2 of this review. Other environmental conditions also influence mucus swelling. A low pH, for example, will reduce the abundance of charges on the expanding polyanionic macromolecules and reduce the mutual repulsion thought to drive the swelling. These factors are interant within the Donnan-system which underpins ion exchange theory and have been implicated in a range of abnormal physiological systems. Fish eggs exposed to aluminium ions and acid rain may be influenced in several ways. The perivitelline mucilage that surrounds the embryo is a polyanionic macromolecule, retained by the eggs outer membrane. It acts as a physical and possible chemical barrier. Fish eggs exposed to water at low pH, swell less than those exposed to neutral pH (for example, Shephard 1987b).

The implications of reduced swelling by mucins on epithelia unable to provide optimum conditions for expansion have also recently been the subject of concern in human abnormalities. Cystic fibrosis has long been known to be associated with defective ion transport (Widdicombe *et al* 1985, Boucher *et al* 1986). The single defective gene (reviewed by Riordan 1993) codes for a chloride ion-channel with abnormal regulation. The gene is expressed by many tissues but the most severe problems arise from the accumulation of sticky mucus in the airway. Most research has concentrated on the notion that the ion-transport mediated water-movement in the CF airway is inadequate to supply water for hydration of mucus (Widdicombe & Widdicombe 1995). A defective ion channel may, therefore, block the airways ability to regulate the availability of water. Recent research, however, also emphasises other aspects of CFTR abnormality. CFTR is predominantly expressed by goblet cells and secretory gland cells, not by water transporting epithelial cells (for example, Hayden & Carey 1996). Furthermore, respiratory glycoproteins in cystic fibrosis have altered biochemical properties, particularly in relation to their degree of sulphation (Zhang *et al* 1995) indicating a primary intracellular mucus processing defect. Even more revealing is the realisation that the CF defect gives rise to a whole range of defective protein secretory systems which involve exocytosis (e.g. Mills *et al* 1995). The possibility exists that CFTR regulates ion transport into or out of secretory granules, and is, therefore, a key component of mucin exocytosis. Defective regulation of the epithelial micro-environment potentially has severe consequences.

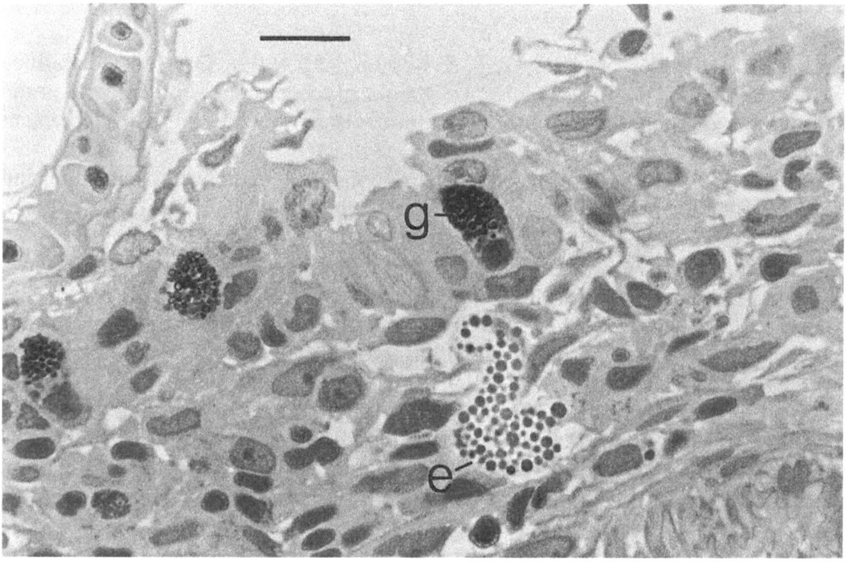

Fig. 3.1. An eosinophilic granulocyte (e) below a goblet cell (g) within the epidermis of a gill filament. The preparation is a semi thin section of glutaraldehyde-fixed, LR gold embedded filament from *Oreochromis mossambicus* viewed with phase contrast optics. Eosinophilic granulocytes in the fish gill possesses many of the characteristics of their mammalian counterparts (Barnett *et al* 1996) and they may have a role in regulating the release of mucus by goblet cells in stressful circumstances. Section by David Alonso. Scale bar =10mm.

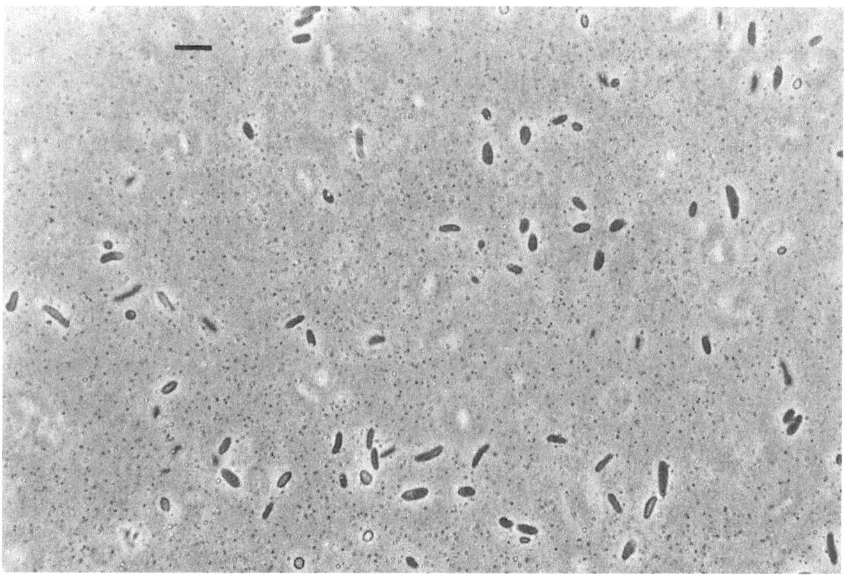

Fig. 3.2. Mucous granules released by the epidermis of the terrestrial slug *Milax budapestensis*. Granules burst on the slug surface, apparently spontaneously, to release mucus which swells by absorbing water. Scale bar =10mm.

References

Adler, K.B., Holden-Stauffer, W.J. & Repine, J.E. 1990. Oxygen metabolites stimulate release of high-molecular-weight glycoconjugates by cell and organ cultures of rodent respiratory epithelium via an arachidonic acid-dependent mechanism. J. Clin. Invest. 85: 75-85.

Ahuja, S.K. 1970. Chloride cell and mucous cell response to chloride and sulphated enriched media in the gills of *Cambusia affines* and *Catla cata*. J. exp. Biol. 173: 231-50.

Al-Bazzaz, F.J. 1994. regulation of Na and Cl transport in sheep distal airways. Am. J. Physiol. 267: L193-L198.

Barbet, J.P., Chauveau, M., Labbe, S. & Lockhart, A. 1988. Breathing dry air causes acute epithelial damage and inflammation of the guinea pig trachea. J. Appl. Physiol. 64: 1851-1857.

Balm, P.H.M., Vanlieshout, E., Lokate, J. & Bonga, S.E.W. 1995. Bacterial lipopolysaccharide (LPS) and interleukin 1 (IL-1) exert multiple physiological effects in the *Tilapia oreochromis mossambicus* (Teleostei). J. Comp. Physiol. - B, Biochemical, Systemic & Environmental Physiology, 165(2): 85-92.

Barnett, R.R., Akindele, T., Orte, C. & Shephard, K.L. 1996. Eosinophilic granulocytes in the epidermis of tilapia (*Oreochromis mossambicus*) gill filaments studied *in situ*. J. Fish Biol. (in press).

Blackstock, N. & Pickering, A.D. 1982. Changes in the concentration and histochemistry of epidermal mucous cells during the alevin and fry stages of the brown trout, *Salmo trutta*. J. Zool. Lond. 197: 463-471.

Boucher, R.C., Stutts, M.F., Knowles, R.C., Cantley, L. & Gatzy, J.T. 1986. Na transport in cystic fibrosis respiratory epithelia. Abnormal basal rate and response to adenylate cyclase activation. J. Clin. Invest. 78: 1245-1252.

Bouhanna, C., Berthon, B., Combettes, L., Claret, M. & Laboisse, C.L. 1994. Role of calcium in carbachol- and neurotensin-induced mucin exocytosis in a human colonic goblet cell line and cross-talk with the cyclic AMP pathway.

Breckenridge, L.J. & Almers, W. 1987. Final steps in exocytosis observed in cells with giant secretory granules. Proc. Natl. Acad. Sci. USA 84: 1945-9.

Burden, C.E. 1956. The failure of hypophysectomized *Fundulus heteroclitus* to survive in freshwater. Biol. Bull. mar. biol. Lat. Woods Hole 110: 8-28.

Burnstock, G. 1986a. Purines as cotransmitters in adrenergic and cholinergic neurones. Prog. Brain Res. 68: 193-203.

Burnstock, G. 1986b. The non-adrenergic non-cholinergic nervous system. Arch. Int. Pharmacodyn Terh. 280, 1-15.

Burton, D. & Everard, B.A. 1991. The effect of androgen treatment on the epidermis of post-spawned winter flounder *Pseudopleuronectes americanus* (Walbaum). J. Fish Biol. 38: 73-80.

Carlstedt, I. & Sheehan, J.K. 1989. Structure and macromolecular properties of cervical mucus glycoproteins. Symp. Soc. Exp. Biol. 43: 289-316.

Cole, P. 1982. Modification of inspired air. In: *The Nose*, edited by D.F. Proctor & I.B. Anderson, Amsterdam: Elsevier: 351-376.

Davis, C.W., Dowell, M.L., Lethem, M. & Van Scott, M. 1992. Goblet cell degranulation in isolated canine tracheal epithelium: response to exogenous ATP, ADP, and adenosine. Am. J. Physiol 262 (Cell Physiol. 31): C1313-C1323.

Daviskas, E., Gonda, I. & Anderson, S.D. 1991. Local airway heat and water vapour losses. Respir. Physiol. 84: 115-132.

De Ruiter, A.J.H. & Mein, C.G. 1982. Testosterone-dependent transformation of nephronic tubule cells into serous and mucous gland cells in stickleback kidneys in vivo and in vitro. Gen. comp. Endocrinol. 47: 70-83.

Deyrup-Olsen, I., Luchtel, D.L. & Martin, A.W. 1983. Components of mucus of terrestrial slugs (Gastropoda). Am. J. Physiol. 245: R448-R452.

Fernandez, J.M., Villalon, M. & Verdugo, P. 1991. Reversible condensation of mast cell secretory products *in vitro*. Biophys. J. 59: 1022-7.

Fletcher, T.C. & White, A. 1973. Lysozyme activity in the plaice (*Pleuronectes platessa*). Experientia 29: 1283-5.

Frizzel, R.A. 1988. Role of absorptive, and secretory processes in hydration of the irway surface. Am. Rev. Respir. Dis. Suppl. 138: S3-S6.

Handy, R.D. & Eddy, F.B. 1991. The absence of mucus on the secondary lamellae of unstressed rainbow trout, *Oncorhyncus mykiss* (Walbaum). J. Fish Biol. 38: 153-5.

Hayden, U.L. & Carey, H.V. 1996. Cellular localisation of cystic fibrosis transmembrane regulator protein in pigment and mouse intestine. Cell & Tiss. Res. 283(2): 209-213.

Hayes, J.P., Kuo, H.P., Rohde, J.A.L., Taylor, A.J.N., Barnes, P.J., Chung, K.F. & Rogers, D.F. 1995. Neurogenic goblet cell secretion and bronchoconstriction in guinea pigs sensitised to trimellitic anhydride. Eur. J. Pharmacol. 292(2): 127-134.

Helfferich, F. 1962. *Ion exchange*. New York: McGraw-Hill, 482pp.

Hoffstein, S.T., Malo, P.E., Bugelski, P. & Wheeldon, E.B. 1990. Leukotriene D_4 (LTD_4) induces mucus secretion from goblet cells in the guinea pig respiratory epithelium. Exp. Lung Res. 16: 711-725.

Hootman, S.R. & Deondarza, J. 1995. Regulation of goblet cell degranulation in isolated pancreatic ducts. Am. J. Physiol. 31(1): G24-G

Hughes, G.M. & Wright, D.E. 1970. A comparative study of the ultrastructure of the water-blood pathway in the secondary lamellae of teleost and elasmobranch fishes - benthic forms. Z. Zellforsch. mikrosk. Anat. 104: 478-93.

Iger, Y., Balm, P.H.M. & Bonga, S.E.W. 1994. Cellular responses of the skin and changes in plasma cortisol levels of trout (*Oncorhynchus mykiss*) exposed to acidified water. Cell & Tissue Res. 278(3): 535-542.

Iger, Y., Balm, P.H.M., Jenner, H.A. & Bonga, S.E.W. 1995. Cortisol induces stress-related changes in the skin of Rainbow trout (*Oncorhynchus mykiss*). Gen. & Comp. End. 97(2): 188-198.

Kapeleta, M.V., Jimenez-Mallebrera, C., Carnicer-Rodriguez, M.J., Cook. A. & Shephard, K.L. 1996. Production of mucous granules by the terrestrial slug *Arion ater* L. J. Moll. Stud. 62: 251-256.

Kilburn, K.H. 1968. A hypothesis for pulmonary clearance and its implications. Am. rev. Respir. Dis. 98: 449-463.

Klinger, J.D., Tandler, B., Liedtke, C.M. & Boat, T.F. 1984. Proteinases of *Pseudomonas aeruginosa* evoke mucin release by tracheal epithelium. J. Clin. Invest. 74: 1669-1678.

Laurent, P. 1984. Gill internal morphology. In: Hoar, W.S. & Randall, D.J., eds. Fish Physiology, Vol. XA. Orlando FL: Academic Press: 73-183.

Maggi, C.A. & Mell, A. 1987. The sensory-efferent function of capsaicin-sensitive sensory neurones. Gen. Pharmacol. 19: 1-43.

Man, S.F.P., Adams III, G.K. & Proctor, D.F. 1979. Effects of temperature, relative humidity, and mode of breathing on canine airway secretions. J. Appl. Physiol. 46: 205-210.

Marshall, W.S. 1976. Effects of hypophysectomy and ovine prolactin on the epithelial mucus secreting cells of the pacific staghorn sculpin *Leptocottus armatus* (Teleostei: Cottidae). Can. J. Zool. 54: 1604-9.

Marshall, W.S. 1979. Effects of salinity acclimation, prolactin, growth hormone and cortisol on the mucous cells of *Leptocottus armatus* (Teleostei: Cottidae). Gen. comp. Endocrinol. 37: 358-68.

Martin, A.W. & Deyrup-Olsen, I. 1986. Function of the epithelial channel cells of the body wall of a terrestrial slug, *Ariolimax columbianus*. J. Exp. Biol. 121: 301-314.

Mattheij, J.A.M. & Stroband, H.W.J. 1971. The effects of osmotic experiments and prolactin on the mucous cells in the skin and the ionocytes in the gills of the teleost *Chichlasoma biocellatum*. Z. Zellforsch 121: 93-101.

McCahon, C.P., Pascoe, D. & McKavanagh, C. 1987. Histochemical observations on the salmonids *Salmo salar* L. and *Salmo trutta* L. and the ephemeropterans *Baetis rhodani* (Pict.) and *Ecdyonurus venosus* (Fabr.) following a simulated episode of acidity in an upland stream. Hydrobiologia 153: 3-12.

McDonald, D.G. & Wood, C.M. 1993. Branchial mechanisms of acclimation to metals in freshwater fish. In. Rankin, J.C. & Jensen, F.B. (eds.). *Fish ecophysiology*, Chapman & Hall, London.

McDonald, D.G., Wood, C.M., Rhem, R.G., Mueller, M.E., Mount, D.R. and Bergman, H.L. 1991b. Nature and time course of acclimation to aluminium in juvenile brook trout (*Salvelinus fontinalis*). 1. Physiology. Can. J. Fish aquat. Sci., 48, 2006-15.

McFadden, E.R. 1983. Respiratory heat and water exchange: physiological and clinical implications. J. Appl. Physiol. 54: 331-336.

McFadden, E.R., Pichurko, B.M., Bowman, H.F., Ingenito, E., Burns, S., Dowling, N. & Solway, J. 1985. Thermal mapping of the airways in humans. J. Appl. Physiol. 58: 564-570.

Merlin, D., Augeron, C., Tien, X.Y., Guo, X., Laboisse, C.L. & Hopfer, U. 1994. ATP-stimulated electrolyte and mucin secretion in the human intestinal goblet cell line HT29-CL.16E. J. Membrane Biol. 137(2): 137-149.

Mills, C.L., Dorin, J.R., Davidson, D.J., Porteus, D.J., Alton, E.W.F.W., Dormer, R.L. and McPherson, M.A. 1995. Decreased beta-adrenergic stimulation of glycoprotein secretion in CF mice submandibular glands - reversal by the methylxanthine, IBMX. Bioch. & Biophys. Res. Comm. 215(2): 674-681.

Mittal, A.K., Fujimori, O., Ueda, H. & Yamada, K. 1995. Carbohydrates in the epidermal mucous cells of a freshwater fish *Mastacembelus pancalus* (Mastacembelidae, Pisces) as studied by electron-microscopic sytochemical methods. Cell Tissue Res. 280: 531-539.

Monck, J.R., Oberhauser, A.F., Alvarez de Toledo, G. & Fernandez, J.M. 1991. Is swelling of the secretory granule matrix the force that dilates the exocytotic fusion pore? Biophys. J. 59: 39-47.

Muniz, I.P. & Leivestad, H. 1980. Toxic effects of aluminium on the brown trout, *Salmo trutta* L. In: Drablos, D. & Tollan, A., eds. *Ecological Impact of Acid Precipitation*. Oslo, Norway, SNSF project: 320-21.

Nathanson, I., Widdicombe, J.H. & Nadel, J.A. 1983. Effects of amphotericin B on ion and fluid movement across dog tracheal epithelium. J. Appl. Physiol. 55: 1257-1261.

Nicholl, C.S. & Bern, H.A. 1971. On the actions of prolactin among the vertebrates: is there a common denominator? In: Wolstenholme, G.E.W. & Knight, J. (eds.) *Ciba Foundation Symposium on Lactogenic Hormones*. London: Churchill-Livingstone: 299-324.

Nilsson, S. 1984. Innervation and pharmacology of the gills. In *Fish Physiology*, Vol. XA (eds. W.S. Hoar & D.J. Randall), Academic Press, Orlando, FL: 185-227.

Ogawa, M. 1970. Effects of prolactin on the epidermal mucous cells of the goldfish, *Carassius auratus* L. Can. J. Zool. 48: 501-3.

O'Grady, P. 1991. Pharmacology and therapeutic applications of the quinoline antibiotic flumequine in salmonid fish. PhD thesis, University College, Galway, National University of Ireland.

Ojha, J. & Munshi, J.S.D. 1974. Histochemical and histophysiological observations on specialised branchial glands of a freshwater mud-eel *Macrognathus aculeatum* (Bloch) (Mastacembelidae: Pisces). Mikroskopie 30: 1-16.

Olivereau, M. & Lemoine, A.M. 1971. Action de la prolactire cheq l'anguille intacte et hypophysecsomisee. Z. vergl. Physiol. 73: 34-43.

Olver, R.E., Davis, B., Marin, M.G. & Nadel, J.A. 1975. Active transport of Na^+ and Cl^- across the canine tracheal epithelium *in vivo*. Am. Rev. Respir. Dis. 112: 811-815.

Peatfield, A.C. & Richardson, P.S. 1983. Evidence for non-cholinergic, non-adrenergic nervous control of mucus secretion into the cat trachea. J. Physiol. 342: 335-345.

Pickering, A.D. 1976. Synthesis of an acetyle neuraminic acid from ^{14}C glucose by the epidermis of the brown trout, *Salmo trutta* L. Comp. Biochem. Physiol. 54B: 325-8.

Pickering, A.D. & Macey, D.J. 1977. Structure histochemistry and the effect of handling on the mucous cells of the epidermis of the char *Salvelinus alpinus* (L.). J. Fish Biol. 10: 505-12.

Pickering, A.D. & Richards, R.H. 1980. Factors influencing the structure, function and biota of the salmonid epidermis. Proc. R. Soc. Edin. 79B: 93-104.

Powell, M.D., Wright, G.M. & Burka, J.F. 1990. Eosinophilic granule cells in the gills of rainbow trout, *Oncorhynchus mykiss*: evidence of migration? J. Fish Biol. 37: 495-497.

Rahmoune, H. & Shephard, K. 1994. Regulation of airway surface liquid on the isolated guinea pig trachea. Pulmonary Pharmacology 7: 265-269.

Rahmoune, H. & Shephard, K. 1995. State of airway surface liquid on guinea pig trachea. J. Appl. Physiol. 78(6): 2020-2024.

Rammarine, S.I., Hirayama, Y., Barnes, P.J. & Rogers, D.F. 1994. 'Sensory-efferent' neural control of mucus secretion: characterisation using tachykinin receptor antagonists in ferret trachea *in vitro*. Brit. J. Pharmacol. 113: 1183-1190.

Raphael, G.D., Meredith, S.D., Baraniuk, H.N., Druce, H.M., Banks, S.M. & Kaliner, M.A. 1989. The pathophysiology of rhinitis. II. Assessment of he sources of protein in histamine induced nasal secretions. Am. Rev. Respir, Dis. 139: 791-800.

Riordan, J.R. 1993. The cystic fibrosis transmembrane conductance regulator. Annu. Rev. Physiol. 55: 609-30.

Roberts, R.J. & Bullock, A.M. 1980. The skin surface ecosystem of teleost fishes. Proc. R. Soc. Edinb. u: 87-91.

Rogers, D.F. 1994. Airway goblet cells - responsive and adaptable front-line defenders. European Resp. J. 7(9): 1690-1706.

Rogers, D.F. 1995. Neurokinin receptors subserving airways secretion. Can. J. Physiol. & Pharmacol. 73(7): 932-939.

Shephard, K.L. 1987a. Ion-exchange phenomena regulate the environment of embryos in the eggs of freshwater fish. Comp. Biochem. Physiol. 88A: 659-62.

Shephard, K.L. 1987b. The influence of water pH on the perivitelline fluid of perch (*Perca fluviatilis*) eggs. Comp. Biochem. Physiol. 86C: 383-6.

Shephard, K.L. 1992a. Studies on the fish-gill microclimate: interactions between gill tissue, mucus and water quality. Env. Biol. Fishes 34: 409-20.

Shephard, K.L. 1992b. Studies on the fish-gill microclimate. J. comp. Physiol. 162B: 231-40.

Shephard, K.L. 1994. Functions for fish mucus. Rev. Fish Biol & Fisheries 4: 401-429.

Shephard, K.L. & Rahmoune, H. 1994. Evaporation-induced changes in airway surface liquid on an isolated guinea pig trachea. J. Appl. Physiol. 76(3): 1156-1165.

Stuart, T.A. 1953. Spawning migration reproduction and young stages of loch trout (*Salmo trutta* L.). Freshwat. Salm. Fish Res. 5: 1-39.

Tanaka, T. 1981. Gels. Sci. Am. 244: 124-38.

Tanaka, Y., Yoshida, Y. & Hirano, M. 1995. Precise localisation of VIP-, NPY- and TH-immunoreactivities of cat laryngeal glands. Brain Res. Bull. 36(3): 219-224.

Uvnas, B. & Aborg, C.H. 1977. On the cation exchanger properties or rat mast cell granule and their storage of histamine. Acta. Physiol. Scand. 100: 309-14.

Verburgvankemenade, B.M., Weyts, F.A.A., Debets, R. & Flik, G. 1995. Carp macrophages and neutrophilic granulocytes secrete an interleukin-1-like factor. Dev. & Comp. Imm. 19(1): 59-70.

Verdugo, P. 1990. Goblet cell secretion. Ann. Rev. Physiol. 52: 157-176.

Verdugo, P. 1991. Mucus exocytosis. Am. Rev. of Resp. Dis. 144: S33-S37.

Virabhadrachari, V. 1961. Structural changes in the gills intestine and kidney of *Etroplus maculatus* (Teleostei) adapted to different salinities. Q. J. microsc. Sci. 102: 361-9.

Welsh, M.J., Widdicombe, J.H. & Nadel, J.A. 1980. Fluid transport across the canine tracheal epithelium. J. Appl. Physiol. 49: 905-909.

Wendelaar Bonga, S.E. 1978. The effects of changes in external sodium, calcium and magnesium concentrations on prolactin cells, skin and plasma electrolytes of *Gasterosteus aculeatus*. Gen. comp. Endocrinol. 34: 265-75.

Wendelaar Bonga, S.E. & Meis, S. 1981. Effects of external osmolarity, calcium and prolactin on growth and differentiation of the epidermal cells of the cichlid teleost, *Sarotherodon mossambicus*. Cell Tiss. Res. 221: 109-23.

Westfall, B.A. 1945. Coagulation film anoxia in fishes. Ecology 26: 283-7.

Whitear, M. 1986. The skin of fishes including cyclostomes. In: Bereiter-Hahn, J., Matottsy, A.G. & Richards, K.S., eds. Biology of the Integument, Vol. 2. Berlin: Springer, pp. 8-64.

Whitear, M. 1970. The skin surface of bony fishes. J. Zool., Lond. 160: 437-54.

Widdicombe, J.H., Welsh, M.J. & Finkbeiner, W.E. 1985. Cystic fibrosis decreases the apical membrane chloride permeability of monolayers cultured from cells of tracheal epithelium. Proc. Nat. Acad. Sci. USA 82: 6167-6171.

Widdicombe, J.H. & Widdicombe, J.G. 1995. Regulation of human airway surface liquid. Resp. Physiol. 99: 3-12.

Zaccone, G. 1981. Effect of osmotic stress on the chloride and mucous cells in the gill epithelium of the freshwater teleost, *Barbus filamentosus* (Cypiriniformes, Pisces). A structural and histochemical study. Acta Histochem. 68: 147-159.

Zhang, Y.L., Doranz, B., Yankaskas, J.R. & Engelhardt, J.F. 1995. Genotypic analysis of respiratory mucous sulfation defects in cystic fibrosis. J. Clin. Inv. 96(6): 2997-3004.

Zimmerberg, J., Curran, M., Cohen, F.S. & Brodwick, M. 1987. Simultaneous electrical and optical measurements show that membrane fusion precedes secretory granule during exocytosis of beige mouse mast cells. Proc. Nat. Acad. Sci. USA 84: 1585-9.

Osmotic and Ionic Regulation in Cyclostomes

J. C. Rankin

Biologisk Institut, Odense Universitet, 5230 Odense M, Denmark

4.1. Introduction

Lampreys and hagfish were traditionally grouped together as cyclostomes ("round mouths") in spite of the many differences in their anatomy, physiology and biochemistry (Hardisty, 1979), but consideration of their osmoregulatory mechanisms, or lack of them, has long been recognized as an indication of the ancient divergence of the two groups (Potts & Parry, 1964).

500 to 400 million years ago there was a great radiation of vertebrates into a variety of groups of jawless fish possessing dermal skeletons of scales and/or bony plates. Most of these became extinct before the beginning of the Carboniferous period, but three lines survived and are represented today by the hagfishes, the lampreys and the gnathostomes, or jawed vertebrates (Forey and Janvier, 1993). In contrast to the latter, the first two groups have changed little since the Carboniferous and comparisons between the three may tell us something about basic physiological mechanisms common to all vertebrates.

However, very few species of hagfish and lampreys have survived and, lacking jaws, they seem only to have been able to compete with the most successful group of vertebrates (in terms of number of species), the teleost fish, by adapting to a few specialised ecological niches. There is, therefore, always room for speculation about which are primitive features and which secondary adaptations, particularly with physiological mechanisms which must adapt, in both the evolutionary and physiological senses of the word, to meet current circumstances.

4.2. Hagfish

Hagfish do not osmoregulate, i.e. possess the ability to maintain an osmotic gradient between their internal and external environments, but they do have limited powers of ionic and volume regulation. In common with many marine invertebrates they regulate magnesium and sulphate ions at below the sea water concentrations (see for example data for the Atlantic hagfish, *Myxine glutinosa* and a range of invertebrate groups in Potts & Parry, 1964, Table III.2).

It has been suggested that the liver is the site of magnesium excretion, since Rall & Burger (1967) found that the urine/plasma magnesium ratio was close to 1 in four fish studied but the bile/plasma ration was over 4, biliary concentrations being almost twice those of sea water. This contrasts with all other vertebrates in which magnesium is regulated renally; this may be related to the lack of renal tubules (apart from short neck segments) in hagfish. The archinephric duct does, however, exhibit some of the functions of the renal tubules e.g. glucose reabsorption (Fels, Raguse-Degener & Stolte, 1989) and the studies of Morris (1965) did suggest a role for the kidney in magnesium excretion.

4.2.1 Volume Regulation

Hagfish have a small number (about 70 in *Myxine*) of very large, segmentally-arranged glomeruli and the observation of swollen archinephric ducts after exposure to slightly diluted (by about 6%) sea water led to the suggestion that they might be involved in volume regulation (Fänge, 1963), possibly acting as a valve with increasing filtration if body fluid volume increases and vice versa (Rall & Burger, 1967). Whilst glomerular capillary pressure in anaesthetised *Eptatretus stouti* was insufficient for ultrafiltration to occur, so some as yet unknown mechanism must be involved in urine formation under these circumstances (Riegel, 1986), in *in situ* perfused glomeruli filtration rate rose steeply at segmental arterial pressures above about 3 mm Hg (Riegel, 1978).

In contrast to the anaesthetised animals which had a dorsal aortic pressure of 6.8 mm Hg (Riegel, 1986), swimming *Eptatretus* had a dorsal aortic pressure of 11.6 mm Hg (Forster et al, 1988). Intravenous sea water injection increases arterial pressure in *Eptatretus* (Chapman et al, 1963) and in perfused hearts power output increased as input pressure was raised in the range 2 - 3 mm Hg, although it fell if dorsal aortic pressure exceeded 15mm Hg. Cardiac output rose with increasing input pressure until it reached 4 mm Hg, then levelling off (Forster, 1989). One mechanism thus exists for an increased body fluid volume to increase glomerular filtration via arterial pressure; another is that there was a much greater effect of increasing arterial pressure if venous pressure was increased by around 4 mm Hg in *Myxine* (Stolte & Schmidt-Nielsen, 1978).

An organ capable of volume regulation would seem to have been an essential pre-adaptation to colonisation of freshwater, as in its absence a marine animal would not survive for more than a few hours after entering an estuary. Interestingly, the Protochordate *Branchiostoma* (previously known as *Amphioxus*) appears incapable of volume regulation, possessing nephridia instead of glomeruli (Binyon, 1979). Binyon concluded that "The 'protovertebrate' would therefore have required a rather different apparatus of urine production in order to make the transition to estuarine and finally to fresh water". (Presuming a freshwater origin of all vertebrates, but see section 4.3)

Although hagfish do not osmoregulate, knowledge of their volume regulatory mechanism, if that indeed is what the kidney is, allows of interesting comparisons with lamprey and gnathostome kidneys. Further information on hagfish kidneys will be contained in the forthcoming book "The Biology of Hagfishes" (eds. Jørgensen, Weber, Lomholt & Malte; in preparation).

4.3. Lampreys

Present day lampreys live in freshwater or are anadromous migrants. The eggs hatch into filter-feeding ammocoete larvae, which metamorphose into resident adults or downstream migrants. The most studied species, the river lamprey, *Lampetra fluviatilis* and the sea lamprey, *Petromyzon marinus* migrate to sea water (except for the landlocked population of the latter in the Great Lakes) where they feed by attaching to other fish with their sucker, rasping through the skin and sucking the prey's blood. On return to freshwater they do not feed, the gonads growing at the expense of other tissues and they die after spawning (Larsen & Dufour, 1993).

The ancestors of lampreys and gnathostomes are presumed to have lived in freshwater at some stage during their evolution. Increased numbers of glomeruli are associated with the necessity to eliminate large branchial water influxes and lowered plasma ionic concentrations, compared to hagfish, reduce the osmotic gradient into the body. This causes osmotic problems in species which have re-colonised sea water or are euryhaline. Marine members of the class most closely related to lampreys, the elasmobranchs, have become isotonic with sea water by retention of the organic osmolytes urea and trimethylamine oxide, but lampreys hypo-osmoregulate in the same way as teleosts, drinking seawater and eliminating the excess sodium and chloride ions branchially (Potts & Parry, 1964).

Large numbers of glomeruli are clearly associated with freshwater osmoregulation. Many marine teleosts have smaller glomeruli and in a number of cases they have been lost altogether, resulting in aglomerular kidneys which produce urine by tubular secretion. It does not seem conceivable that hagfish have evolved a reduced number of glomeruli in this way, as they have retained the primitive segmental arrangement of the glomeruli which has been lost in lampreys and gnathostomes. It is much more reasonable to suppose that the latter groups, but not the line which gave rise to the hagfish, had freshwater ancestors.

4.3.1. Osmoregulation in Lampreys

A considerable body of early work on lamprey osmoregulation is summarised in Morris (1972); however it consists almost exclusively of studies on osmoregulation in freshwater, as data on lampreys caught at sea came from only 2 individuals at that time. Lampreys osmoregulate in a similar manner to teleosts in freshwater. Water entering osmotically across the gills is eliminated renally, active tubular reabsorption leading to a dilute urine. The small quantity of ions which are lost in the urine and by diffusion across the gills are replaced by active uptake at the gills (Pickering, 1973), presumably by the mitochondria-rich cells present in the interplatelet area (Morris & Pickering, 1975).

Downstream migrants, following metamorphosis, can be successfully acclimated to full sea water, developing exposed chloride cells in the gills with many similarities to those of teleosts (Pickering and Morris, 1976), but also structural (Bartels and Potter, 1991) and ulstrastructural (Barter et al, 1993) differences but they are too small for many physiological experiments and adults re-entering freshwater rapidly lose their marine osmoregulatory abilities (for earlier references see Beamish, 1980). As it is very difficult to catch lampreys at sea, many studies have used upstream migrants acclimated to high salinities for short periods (Abou-Seedo & Potter, 1979) or to diluted (e.g. 50%) sea water (Morris, 1958). The following causes of the loss of marine osmoregulatory capability have been adduced: increased permeability to water; breakdown of the swallowing mechanism; reduced ability to absorb monovalent ions and water from the ingested 50% sea water; and a loss in the large mitochondria-rich "chloride output cells" of the gills (Pickering & Morris, 1970).

Lampreys are caught in coastal eel traps before entering freshwater, whilst they are still capable of long term osmoregulation in full sea water. A few fish caught in the Humber Estuary were used to study kidney function and found to be able to produce urine hyperosmotic to plasma (Logan, Morris & Rankin, 1980). After many years during which it proved impossible to obtain any lampreys, a new source was found enabling continuance of studies on renal and other aspects of marine osmotic and ionic regulation.

4.3.2 The Lamprey Kidney

Lamprey kidneys are completely different to those of hagfish. In fact they have the same number of glomeruli in relation to body weight as human kidneys. This coincidence should help us to remember that lampreys are more closely related to humans than they are to hagfishes! (Forey and Janvier, 1993). Another coincidence is the similarity in the tubules. Fig. 4.1 shows a drawing of a cast of a Microfil injected lamprey nephron. An *in situ* photograph of the cast, taken after the kidney had been dehydrated in alcohols and cleared in methylsalicylate (Logan, 1978) was scanned into Adobe Illustrator 6.0. The drawn-in outlines thus represent exactly the *in situ* proportions. The glomerulus (in the common inaccurate usage of the word - the cast is of course of the Bowman's space which surrounds the glomerulus) is connected by a short ciliated neck segment to the proximal convoluted tubule which leads to a hairpin loop, the distal tubule (not visible on the surface) continuing to the collecting duct which drains into the urinary duct near the tip of the loop (Logan, Moriarty and Rankin, 1980). The urinary duct, which would run across the page over (ventral to *in vivo*) the loop serves to store urine, as lampreys have no urinary bladder.

The evolution of the vertebrate kidney is often thought of as a progression from teleost-type to mammalian-type nephrons (with kidneys of the most recently evolved Class, the birds, commonly described as having a mixture of "reptilian-type" and "mammalian-type") but a characteristic of teleostean kidney evolution has been reduction in, or loss of, nephron elements. Marine teleosts have commonly lost the distal tubule ("diluting segment") and have reduced the glomeruli or even lost them totally to give aglomerular kidneys consisting solely of blind-ending secretory proximal tubules. The lampreys and their nearest relatives the elasmobranchs have nephrons as or more complex than those of mammals (elasmobranchs having 2 hairpin loops per nephron!, Hentschel, 1987). Although they could well have arisen by parallel evolution, it must be emphasised that hairpin loops are more common than is generally realised in vertebrate kidneys.

Fig. 4.1. Drawing of a Microfil cast of a *Lampetra fluviatilis* nephron. The tip of the loop is at the outer (lateral) margin of the kidney ca. 2mm from the glomerulus) For further details see text.

In a lamprey kidney all the glomeruli are usually enclosed in a single tough oblong-shaped capsule running the length of the kidney - the glomus (Youson, 1981) but within this the Bowman's spaces (formed by the continuation of the tubular cells to surround and envelop the glomerulus) can be seen to be separate by Microfil injection into the tubules. Individual glomeruli (capillary loops arising from a single afferent arteriole) can be seen in casts made by intra-arterial injection of Microfil (Logan, Moriarty, Morris and Rankin, 1980). The single glomus has caused confusion in the past, but at least in *Lampetra fluviatilis* there appears to be one tubule draining from each of the corpuscles. Proximal ends of the nephrons appear to be arranged so that adjacent nephrons share glomerular capillaries, but this interpretation is based only on examination of histological sections (Hentschel and Elger, 1987).

The nephron units, as depicted in Fig. 4.1, are arranged in a regular parallel fashion along the length of the kidneys on both sides of the row of glomeruli and most parts, plus the associated blood vessels, are accessible to micropuncture. This, and the fact that river lamprey kidneys respond appropriately to changes in environmental salinity during long periods of anaesthesia (for an example of this see Fig. 4.2), led to the initiation of a long programme of research into lamprey renal mechanisms at the University of Wales, Bangor. Before summarising this work and attempting to integrate it into current research on marine osmoregulation, brief details of materials and methods will be given.

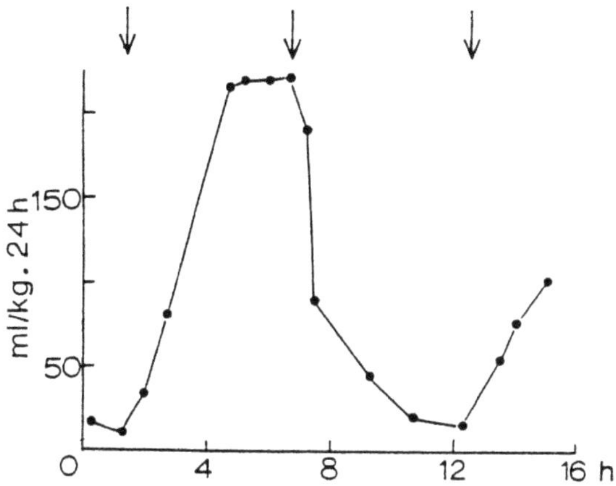

Fig. 4.2. Urine flow rate (ml per kg and 24h) over a 16h period in an anaesthetised river lamprey. The experiment started with approximately iso-osmotic, to use the less euphonius term, 30% sea water bathing the head end of the animal, including the gills. At successive arrows this was replaced by freshwater, then 30% sea water, then freshwater. (From Dalgleish, 1975, unpublished Honours Project Report).

4.3.3 Materials and Methods

Original data presented here comes from lampreys caught by eel fishermen in Ringkøbing Fjord on the west coast of Jutland. Ringkøbing Fjord is a large (ca. 30 x 10 km) brackish lagoon connected to the sea by a regulated channel at the port of Hvide Sande. Osmolality is very variable, up to 440 mOsm kg^{-1} in samples taken at the site of capture. River lampreys were caught in September and October, transported to Odense and kept in either freshwater or in local sea water (usually around 500 mOsm kg^{-1}), the salinity of which was increased as required by the addition of North Sea water. All experiments were carried out at 10°C.

Blood samples were taken under anaesthesia (3-aminobenzoic acid ester methanesulfonate salt, Sigma: 0.065g/l) from a superficial ventral vein in the tail or, if more blood was required, from the caudal vein using a heparinized (ammonium heparin, Leo Pharmaceutical Products, Denmark) syringe. The syringe was either weighed before and after receiving ca. 20μl distilled water containing ca. 40 IU heparin or was allowed to dry overnight so that dilution due to the heparin could be corrected for if required. (0.2ml blood containing 20μl heparin solution would be diluted by 10%!, but usually more blood was taken.)

Urine samples were obtained from fish anaesthetised initially at the above concentration but with periodic adjustments to maintain strong breathing, thus preventing any reduction in urine flow rate (McVicar & Rankin, 1983). The urinary papilla was catheterised with polythene tubing (Portex, UK: 0.7mm int. dia.) with a slight constriction near the tip to allow firm ligaturing into the papilla. A short length of catheter was lightly tied into the papilla and the body cavity gently "milked" to empty the urinary ducts into the air-filled catheter. The area around the papilla was dried with tissue (to avoid contamination with rectal fluid which was iso-osmotic with plasma; there was no risk of contamination with sea water as only the head end of the animal was immersed) and the catheter was removed and the contents immediately expelled into liquid paraffin and analysed as soon as possible. The catheter was then re-ligatured into the papilla and left there for several hours, the lamprey remaining anaesthetised. The "milking" procedure was then repeated exactly as before, with the end of the catheter under liquid paraffin, into which the final catheter contents were expelled. The volume of the droplet was measured to 0.02μl by sucking it into a micropipette (Gilson Microman) using the volume adjustment knob, or by weighing the catheter full and empty if the urine only occupied part of its length. Urine production rate could then be calculated.

Drinking rates were measured by adding ^{51}Cr-EDTA (Amersham, UK) to small tanks (ca. 10l) containing at most 6 lampreys. After 3-5h they were removed to a rinsing bath for 15-20 min and then anaesthetised (0.1 g/l) and counted in a whole body gamma counter (Labtech: Bicron Corp., USA). Counts from aliquots of bath water, corrected for difference in counting geometry, were used to calculate number of counts in the body. Blood samples counted in a higher efficiency counter (1480 Wallac Wizard, Finland) did not give significantly higher counts than background; a number of freshwater-acclimated lampreys also gave background counts and experiments with different immersion times showed that lampreys did not start losing counts anally for about 8h. so accurate calculations of the amount of water swallowed could be made. Lampreys in one group were anaesthetised, as for urine collection, before the ^{51}Cr-EDTA was added to the bath. To estimate absorption of water and salts by the gut, samples of rectal fluid were taken after 24h immersion in ^{51}Cr-EDTA solution.

10μl aliquots of plasma, urine or rectal fluid were taken with a positive-displacement micropipette (Gilson Microman) for osmolality measurement (Wescor 5500 Vapour Pressure Osmometer, USA). Great care is obviously required to prevent dilution of plasma or evaporation of urine if hyperosmotic urine formation by seawater-acclimated lampreys is to be demonstrated. Analysis for inorganic anions and cations was carried out on samples diluted 100 or 200 times (5μl + 995μl or 10μl + 990μl double distilled water) using an ion chromatograph (Dionex 4500I, USA) with anion and cation exchange columns. Linear 3 point calibrations were carried out before every 20 or so samples and were considered acceptable if the r^2 value was >0.999.

4.3.4 Freshwater Osmoregulation

Lampreys have slightly lower plasma osmolality and ionic concentrations than most other vertebrates, a feature they share with many amphibians. This is not due to limited ability to regulate during the upstream migration, as they also maintain these "low" values in hyperosmotic solutions. For example lampreys from Ringkøbing Fjord acclimated to 700mOsm kg^{-1} sea water maintained a plasma osmolality of 271 ± 5.0 mOsm kg^{-1} (mean ± SEM, n = 14) and sodium concentration of 114 ± 1.8 mmol l^{-1} (n = 14) compared to 263.4 ± 2.6mOsm kg^{-1} (n = 10) and 122 mmol l^{-1} (n = 8) for those maintained in freshwater.

The kidney of the freshwater river lamprey regulates water balance in the face of a massive osmotic influx across the gills, urine flow rate being proportional to the osmotic gradient (McVicar and Rankin, 1983) with most of the filtered salts being reabsorbed in hypo-osmotic media giving a urinary sodium concentration of 7.3 mmol l^{-1} in freshwater and 9.5 mmol l^{-1} in 20% sea water. The sodium and chloride ions are reabsorbed in the distal tubule, making the functional assumption that this starts at the tip of the nephron loop, the proximal tubule reabsorbing no more than 10% of the filtered water and salts (Logan, Moriarty and Rankin 1980).

Urine flow rate in hypo-osmotic media is determined by glomerular filtration rate (GFR) which in turn is determined by single nephron glomerular filtration rate (SNGFR) (Moriarty et al, 1978). This is in contrast to the situation in teleosts where urine flow rate and GFR are determined by the number of filtering glomeruli (Brown et al, 1993); the presence of significant numbers of non-filtering glomeruli, as occurs in teleosts(Brown et al, 1978), was ruled out by comparing numbers of filtering glomeruli calculated by dividing GFR by SNGFR with the total number of glomeruli calculated from anatomical observations (Brown et al., 1993). Using a modification of the Hanssen ferrocyanide technique (Brown et al, 1978) out of 232 glomeruli examined from 6 freshwater lampreys 231 were found to have been filtering (Brown and Rankin, in preparation). As lampreys have no renal portal system, all blood perfusing the tubules comes from the glomeruli, whereas teleost tubules can continue normal secretory functions when the glomeruli are shut down or, in the case of aglomerular fish, have disappeared. Hentschel and Elger(1987) suggest that single nephron recruitment would not be possible if there is not a 1:1 correspondence between glomeruli and tubules, but recruitment of small groups of nephrons would still be possible if this were the case.

In freshwater lampreys SNGFR is determined by effective filtration pressure, that is the hydrostatic pressure gradient between the glomerular capillaries and the Bowman's space minus the oncotic pressure due to the plasma proteins (McVicar & Rankin, 1985). The question of what regulates glomerular capillary pressure remains to be resolved.

4.3.5 Marine Osmoregulation

Whilst all Ringkøbing Fjord fish could be placed directly in 500mOsm kg^{-1} (50% sea water) on arrival and then taken up to 700mOsm kg^{-1}, only some could be gradually acclimated to full sea water. These had slightly higher plasma osmolalities than the fish in 700mOsm kg^{-1} but regulated plasma ionic concentrations well (Fig. 4.3). This variation in tolerance of full sea water was investigated and two forms of lamprey were identified. Some, which tended to be smaller, were a shining silver colour. A number of these had been feeding before capture. As the time between capture and transport to Odense was not known (the eel fishermen accumulated them in a cage over a period of several days) the maximum amount of blood in an intestine (16.8% of body weight) in fish sampled on arrival was perhaps of more interest than the mean value (8.7%). A few silver lampreys fed on flatfish in the aquarium but studies during the course of feeding could not be carried out, as the lampreys released their hold and swam away as soon as they were disturbed.

Lampreys were classified subjectively into 5 categories: 1. Silver 2. Half silver 3. Intermediate 4. Half Yellow 5. Yellow or Brown. The silver colour

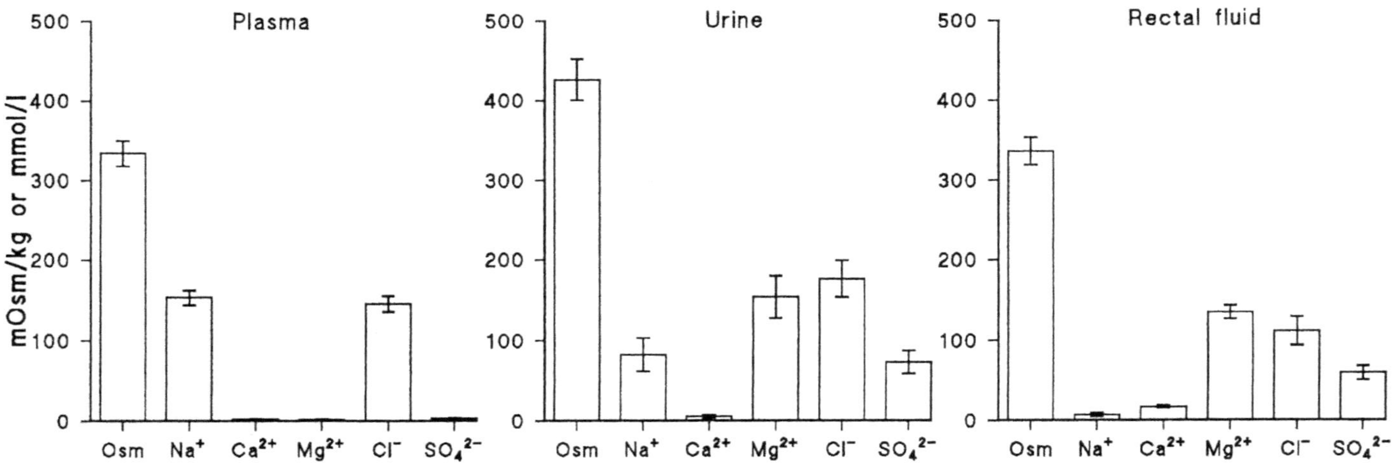

Fig. 4.3 Osmotic and ionic concentrations of plasma, urine and rectal fluid of lampreys acclimated to sea water (composition: 944 mOsm kg; Na^+ 471 mmol 1,1; K^+ 9.0mmol 1 1;Ga^{2+} 10.6mmol 1;1 Mg^{2+} 54.2mmol 1^{-1} ; SO 26.5mmol 1^{-1}).

is presumably the camouflage of the marine form in contrast to the yellow or brown camouflage colouration of the riverine form. In October, when both silver and yellow forms were present, the silver ones had significantly larger intestines and significantly lower gonosomatic ratios (Bendixen, unpublished Bachelor Project Report). Lampreys could be kept in sea water until the normal time of maturation in the spring, with almost 100% survival in running Great Belt surface sea water, which varies between 500 - 700mOsm kg^{-1}. Silver lampreys, although initially having smaller gonads, matured at the normal time, gradually losing their silver coloration. Mature fish had normal gonadosomatic ratios in spite of having been kept throughout the winter in sea water, and even released eggs which shrivelled up in the sea water.

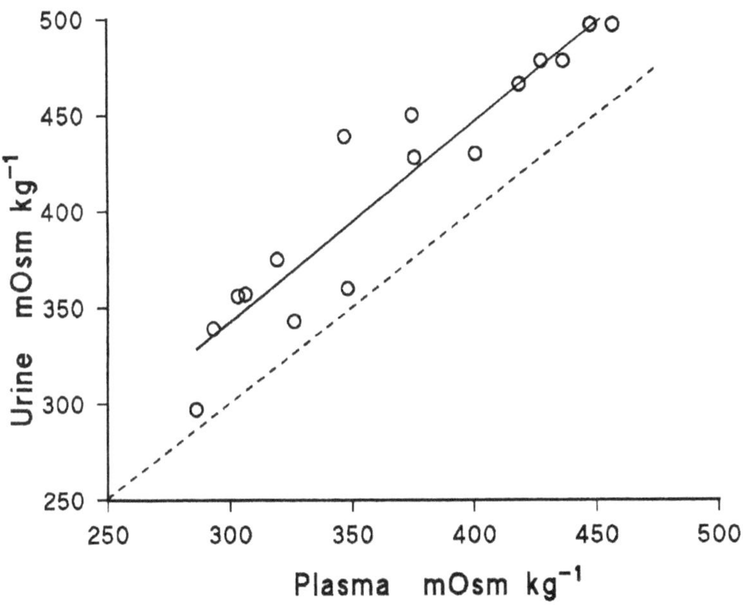

Fig. 4.4 Plasma and urine osmolality in sea water-acclimated lampreys. The solid line is the regression line through the points ($r^2 = 0.89$); the broken line is the iso-osmotic line.

4.3.6 Renal function

The lampreys whose plasma composition is shown in Fig. 4.3 were not categorised by colour, being the ones that survived in North Sea water. Their urine composition is also shown in Fig. 4.3. They produced hyperosmotic urine with very high magnesium and sulphate concentrations. When a range of sea water lampreys, from very good regulators to very bad regulators, was investigated, every one was found to have urine hyperosmotic to plasma (Fig. 4.4) when strict care was taken to avoid erroneous readings (including overcompensation for dilution by heparin in the syringe in some cases where a droplet of heparin solution was observed not to mix with the blood). The mean Urine-Plasma difference was 45.1 ± 5.5mOsm kg^{-1} (mean ± SEM, n = 16) with a maximum of 92, compared to the difference of 80mOsm kg^{-1} found by Logan Morris and Rankin (1980); the ionic concentrations were very similar to their values.

Mean urine flow rate in sea water-acclimated lampreys was 5.76 ml kg^{-1} day^{-1}, lower than the value of 19.5 ml kg^{-1} day^{-1} found by Logan et al, 1980 but a serious problem became apparent when drinking rate measurements were made. Lampreys drink sea water to compensate for branchial osmotic loss, but mild anaesthesia, which did not affect breathing movements, inhibited drinking (Fig. 4.5). As the lampreys were anaesthetised for a number of hours to try to improve the accuracy of the urine flow rate measurements they may, in the absence of drinking, have been becoming progressively dehydrated which may have reduced urine flow. Subsequent experiments showed that plasma osmolality increased during prolonged anaesthesia of sea water-acclimated lampreys (Rankin & Brown, unpublished observations).

SNGFR was not significantly different between lampreys acclimated to 100% (2.9 nl min^{-1}) or 50% (3.1 nl min^{-1}) sea water (Logan, Morris and Rankin, 1980); both values are lower than the freshwater value (7.02 nl min^{-1}; Moriarty et al, 1978) but this is not sufficient to account for the great difference in urine flow rate. The difference is not due to a reduction in the number of filtering glomeruli (Brown and Rankin, in preparation) but to an increase in tubular water reabsorption from 44.4% in freshwater (Logan, Moriarty and Rankin, 1980) to 83.1% in 50% sea water and 89.1% in sea water (Logan Morris and Rankin, 1980). The water reabsorption was predominantly by the distal and/or collecting tubules; less than 10% of filtered water was reabsorbed in the proximal tubule in freshwater or 50% sea water-acclimated lampreys (proximal reabsorption was not measured in sea water fish but would presumably have been low).

The practical problem of collecting micropuncture samples in sea water-acclimated lampreys has so far prevented further investigation of the site of urine concentration and the role of the loop. With an SNGFR of 2.9 nl min^{-1} it is difficult enough collecting proximal tubular fluid, where the volume has not been reduced by fluid reabsorption, but at the end of the nephron, where concentration presumably occurs, fluid will be flowing at only one tenth of this rate. The problem of maintaining water balance during long periods of anaesthesia must also be addressed, but this should be simpler to overcome (e.g. by fluid infusion).

4.3.7 Drinking

Pickering and Morris (1970) measured drinking rate in river-caught *Lampetra fluviatilis* and *Petromyzon marinus* after 24h in 50% sea water . One *Lampetra* did not drink (and lost 14% of its body weight in the day!); drinking rates in the others ranged from 0.12 to 4.15 ml kg^{-1} h^{-1} (with from 9 to 82% of the water swallowed being absorbed). Sea water-acclimated Ringkøbing Fjord lampreys drank at almost twice the maximum rate of Pickering and Morris' lampreys (Fig. 4.5) but had ca. 3 times the osmotic gradient into the body. There was virtually no drinking in freshwater. The sea water rate was at the top end of the range for 12 marine teleost species (1.00 to 7.76 ml kg^{-1} h^{-1} ; mean value 2.7) measured by Perrott et al (1992) but the lampreys had a high branchial tritiated water permeability and the ratio of osmotic to diffusional permeability coefficients was approximately 1.

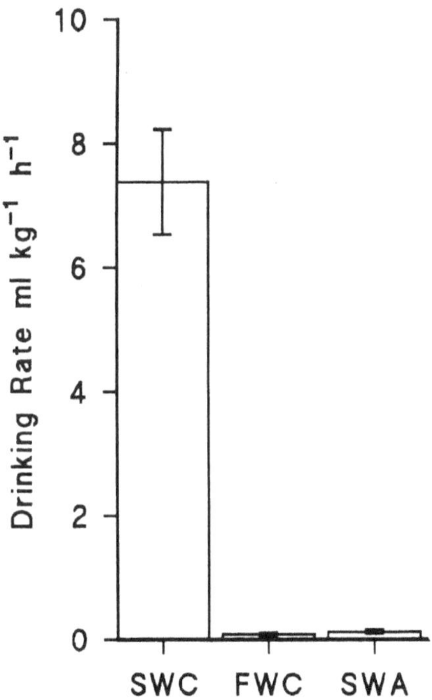

Fig. 4.5. Drinking rates of sea water control (SWC), freshwater control (FWC) and sea water anaesthetised (SWA) lampreys at 10^{0} C.

Samples of rectal fluid taken after 24h in sea water with ^{51}Cr-EDTA showed that 73.6% of the water swallowed had been absorbed. From this, and the ionic concentrations of rectal fluid (Fig. 4.3), it was possible to calculate that 99.3% of the sodium had been absorbed, 89.1% of the potassium and 97.7% of the chloride. Divalent ions could have been removed from the fluid by precipitation as well as by absorption, so the following calculated values for absorption could be overestimates: Magnesium 39.3%; calcium 50.7% and sulphate 68.1%. In any event, divalent ions are not absorbed to the same extent as water.

Precipitate was observed in the rectum; water which is not absorbed must be used to flush this out. This must be taken into account in any experiment to measure drinking rate - too long exposure results in anal loss of marker (Hassan, 1985). The lampreys in Pickering and Morris' (1970) study had their anuses ligated to prevent this happening; they must have been uncomfortable after 24h!

4.3.8 Loss of Marine Osmoregulatory Ability after Freshwater Entry

Ringkøbing Fjord lampreys placed in freshwater immediately on arrival and transferred back to 50% sea water after one month all showed elevated plasma osmolality and ionic concentrations after 3 days. After 7 days plasma calcium, magnesium and sulphate concentrations had all returned to freshwater levels, suggesting that the kidneys were capable of returning to a marine mode of operation but in only about half the fish had the osmotic, sodium and chloride concentrations returned to normal. In these fish gill $Na^+ + K^+$-ATPase activity had increased. In the others, where plasma sodium concentration was greater than 150mmol l^{-1} it had not increased over freshwater levels (Rankin, Madsen and Warne, in preparation). The coloration of these fish was not noted.

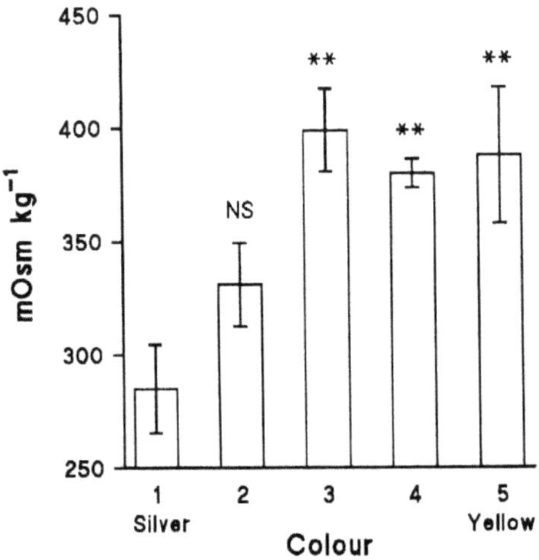

Fig. 4.6 Plasma osmolality 7 days after freshwater (1 months acclimation) to 50% sea water transfer in different groups of lampreys, classified by colour. 1 = Silver 2 = Half silver 3 = Intermediate 4 = Half Yellow 5 = Yellow/Brown Means ± SEMs shown. ** = $p< 0.01$ compared to silver group. NS = not significantly different from silver group. (Tukey protected t test).

When the experiment was repeated ("7 day 50% sea water challenge test" after 1 month in freshwater) silver lampreys were found to have restored their plasma osmolalities to freshwater levels, whereas yellow ones did not (Fig. 4.6). The silver lampreys were also found to have increased branchial $Na^+ + K^+$-ATPase activities whereas the yellow ones did not (Rankin, Madsen and Bendixen, in preparation).

4.4 Control Mechanisms

Almost nothing is known about the endocrine control of lamprey osmoregulation. Partial hypophysectomy (removal of the pro- and meso-adenohypophysis) appeared to have no effect on plasma osmolality and ionic composition or GFR and urine flow rate in freshwater lampreys (Moriarty, 1977) and had no effect on the reduction in urine flow produced by transfer to 10%, 20% or 30% sea water (Cook, 1976, unpublished Honours Project Report).

Lampreys have a single neurohypophysial hormone, arginine vasotocin (AVT) and a single provasotocin gene (Suzuki et al, 1995). AVT is a possible candidate for an antidiuretic hormone controlling urine flow rate, since low doses of AVT are antidiuretic in the teleosts *Anguilla anguilla* (Henderson and Wales, 1974; Babiker and Rankin, 1978), *Oncorhynchus mykiss* (Sinnott and Rankin, 1976) and *Clarias lazera* (Babiker and Rankin, 1980). High doses, which we now know were unphysiological (Balment et al, 1993), were diuretic; this confused the literature for many years. However, in lampreys only diuretic effects are observed (Rankin et al, 1985; Uchiyama and Murakami, 1994; Broadhead and Rankin, in preparation)

There seems to be no difference in plasma AVT concentrations between freshwater and sea water in *Lampetra fluviatilis* (Warne et al, 1993) or freshwater and 50% sea water in *Lampetra japonica*, although the pituitary content was significantly higher in freshwater (Uchiyama et al, 1994) and there was no correlation between plasma osmolality and AVT concentrations during a freshwater to 50% sea water transfer experiment (Warne, Rankin and Balment, unpublished observations).

Low concentrations of AVT induce contractions of lamprey urinary ducts (Rankin et al, 1985; Uchiyama and Murakami, 1994) and could be involved in the regular contractions involved in urine expulsion (McVicar, 1982) and possibly also in sperm expulsion. Lampreys do not have separate urinary and gonadal ducts and it is tempting to speculate that the reproductive and osmoregulatory functions of the two neurohypophysial hormones of higher vertebrates could have originated in a single agnathan hormone acting on the archinephric duct.

If AVT is not the lamprey neurohypophysial hormone, are there other candidates?

Angiotensin II produces glomerular antidiuresis in the perfused trout kidney preparation (Rankin et al, 1984; Dunne and Rankin, 1992), possibly by constricting glomerular afferent arterioles as it does in mammals (Weihprecht et al, 1991). Cyclostomes had been thought to lack a renin-angiotensin system (Nishimura et al, 1970) but lamprey angiotensins have now been found (Takei and Rankin, unpublished observations). Angiotensin II has a dorsal aortic pressor action but produces glomerular antidiuresis in *Lampetra fluviatilis* (Rankin and Broadhead, in preparation).

Many other aspects of the endocrine control of osmoregulation in lampreys remain to be elucidated, including the role of corticosteroids (Weisbart et al, 1980). Nothing is known about the regulation of salt excretion. Infusion of hyperosmotic (400 mOsm kg^{-1}) sodium chloride solution appeared to inhibit sodium and chloride reabsorption by the freshwater lamprey kidney, whereas infusion of equally diuretic quantities of distilled water had no effect on urinary osmolality (Rankin et al, 1983); infusion of 200mmol l^{-1} magnesium sulphate solution produced a diuresis with greatly increased urinary magnesium concentration.

4.5. Conclusions

The emphasis on the kidney in this chapter reflects the amount of work that has been done on lamprey renal physiology; there are many other interesting aspects of lamprey osmoregulation waiting to be investigated. In view of the parallel ways in which lampreys and teleosts have overcome the problems of recolonization of the marine environment (assuming that both had freshwater ancestors) it would be most interesting to measure the transepithelial electrical potentials and branchial ion fluxes of lampreys subjected to different salinities to see if they have a teleost-like mechanism (Potts, Fletcher and Hedges, 1991).

One of the problems of research in lamprey osmoregulation is that the physiological state of the animals is continually changing as they mature, and with the length of time they have been in freshwater. Obtaining more data by repeating an experiment often means having to wait until the following year so it pays to work on a number of aspects in parallel, accumulating data over several seasons.

Lampreys from the Humber Estuary can produce slightly hyperosmotic urine (Logan et al 1980) It was thought possible that they were at an early stage in their upstream migration and may not have had as great a concentrating ability as truly marine feeding animals. However, lampreys from Rinkøbing Fjord, some of which were still feeding, only produce slightly concentrated urine. Although this contradicts the prevailing dogma that only birds and mammals can produce concentrated urine it raises the question: Why do not marine lamprey kidneys produce more concentrated urine to save more scarce free water?

The limited micropuncture data on lampreys acclimated to hyperosmotic media suggests that magnesium and sulphate ions are secreted in the proximal tubules (as they must be in marine teleosts possessing only proximal tubules) and further concentrated by water abstraction in the distal tubules, presumably by solute-linked water reabsorption as the sodium concentration falls from plasma-like concentrations in the proximal tubule to much lower values in the urinary ducts (Rankin et al, 1980). Possibly this solute-linked water movement can only occur against a small osmotic gradient but it is tempting to speculate that the loop/parallel collecting duct system may play a role in urine concentration, in spite of the total lack of evidence, apart from the anatomical similarity to the mammalian arrangement.

Apart from possible mechanistic limitations, the lack of ability to produce very concentrated urine may be due to the nature of the substances concentrated. The mammalian kidney can be very good at concentrating highly soluble salts, such as sodium or potassium chlorides, and urea. The avian kidney excretes less soluble urates which, however, precipitate as small spherical crystals which do not block or damage tubules or ducts, forming (after cloacal/colonic water reabsorption) the familiar white sludge. Marine fish kidneys concentrate magnesium, sulphate, calcium and phosphate ions (the first 3 being present at much higher concentrations in sea water than in blood and lampreys are presumable subjected to quite a phosphate load when digesting blood meals). Youson (1982a) found a white precipitate in the lumina of tubules and urinary ducts of young feeding adult anadromous sea lampreys acclimated to 60% or more sea water, the amount increasing with increasing salt concentrations of the water. He suggested that it might be magnesium sulphate or calcium phosphate (Youson, 1982b) and wavelength-dispersive electron probe X-ray microanalysis of one such crystal (shaped like hydroxyapatite) showed that it contained calcium and phosphorus but not magnesium or sulphur (Rankin, unpublished observation). This was as expected from consideration of the solubility products of the relatively soluble magnesium sulphate but very insoluble calcium phosphates. Hydroxyapatite has been found in eel urine (Butler and Cadinouche, 1995). Other potential precipitates are calcium sulphate and calcium carbonate; even freshwater trout exposed to elevated CO_2 concentrations develop nephrocalcinosis (Smart et al, 1979).

Possibly the reason marine fish do not produce as concentrated urine as mammals may not be due to any innate superiority of the mammalian kidney; rather that the advantages of reducing urine volume further than the 3% of total water loss found here are outweighed by the risk of precipitates blocking or damaging tubules. Lampreys have had much longer than mammals to evolve adequate solutions to the problems they face; their kidneys are not necessarily primitive versions of higher vertebrate kidneys. They still have much to teach us about the evolution of the vertebrate kidney.

4.6. Acknowledgements

This work was supported by the Danish Natural Sciences Research Council. Thanks are due to Lone Morgen for technical assistance. I should like to acknowledge my debt to the many students and colleagues who have collaborated on lamprey investigations and to the authors of the book from which I first learned everything (almost) about osmotic and ionic regulation in animals (Potts and Parry, 1964).

References

Abou-Seedo, F.S. and Potter, I.C. 1979. The estuarine phase in the spawning run of the river lamprey *Lampetra fluviatilis*. J. Zool., Lond. 188: 5-25.

Babiker, M.M. and Rankin, J.C. 1978. Neurohypophysial hormonal control of kidney function in the European eel (*Anguilla anguilla* L.) adapted to seawater and freshwater. J. Endocrinol. 76: 347-358.
Babiker, M.M. and Rankin, J.C. 1980. Diuretic and antidiuretic actions of AVT and isotocin in the teleost fish *Clarias lazera*. Gen. comp. Endocr. 40: 337.
Balment, R.J., Warne, J.M., Tierney, M. and Hazon, N. 1993. Arginine vasotocin and fish osmoregulation. Fish Physiol. Biochem. 11: 189-194.
Bartels, H. and Potter, I.C. 1991. Structure changes in the zonulae occlludents of the chloride cells of young adult lampreys following acclimation to seawater. Cell. Tiss. Res., 265: 447-457.
Baterls, H., Schewe, H. and Poter, I.C. 1993. Structural changes in the apical membrane of lamprey chloride cells after acclimation to seawater. Am. J. Physiol. 265 : C733-739.
Beamish, F.W.H. , 1980. Osmoregulation in juvenile and adult lampreys. Can. J. Fish. aquat. Sci. 37: 1739-1750.
Binyon, J. 1979. *Branchiostoma lanceolatum* - a freshwater reject? J. mar. biol. Ass. U.K. 59: 61-67.
Brown, J.A., Oliver, J.A. and Henderson, I.W. 1978. Single nephron filtration rates (SNGFR) in the trout, *Salmo gairdneri*. Pflügers Arch. 377: 101-108.
Brown, J.A., Rankin, J.C. and Yokota, S.D. 1993. Glomerular haemodynamics and filtration in single nephrons of non-mammalian vertebrates. in Brown, J.A., Balment, R.J. and Rankin, J.C. (eds.) *New insights in vertebrate kidney function*:Cambridge: Cambridge University Press; 1-44.
Butler, D.G. and Cadinouche, M.Z.A. 1995. Hydroxyl-apatite crystals in the urine of stanniectomized fresh-water North-American eels (*Anguilla rostrata LeSueur*). Jap. J. Ichthyol. 42: 7-10.
Chapman, C.B., Jensen, D. and Wildenthal, K. 1963. On circulatory control mechanisms in the Pacific hagfish. Circ. Res. 12: 427-440.
Dunne, J.B. and Rankin, J.C. 1992. Effects of atrial natriuretic peptide and angiotensin II on salt and water excretion by the perfused rainbow trout kidney. J. Physiol. 446: 92P
Fels, L.M., Raguse-Degener, G. and Stolte, H. 1989. The archinephron of *Myxine glutinosa* L. (Cyclostomata). in Kinne, R.K.H. (ed.) *Structure and function of the kidney*: Basel: Karger; 73-102.
Forey, P. and Janvier, P. 1993. Agnathans and the origin of jawed vertebrates. Nature 361: 129-134.
Forster, M.E. 1989. Performance of the heart of the hagfish, *Eptatretus cirrhatus*. Fish Physiol. Biochem. 6: 327-331.
Forster , M.E., Davie, P.S., Davison, W., Satchell, G.H. and Wells, R.M.G. 1988. Blood pressures and heart rates in swimming hagfish. Comp. Biochem. Physiol. 89A: 247-250.
Fänge, R. 1963. Structure and function of the excretory organs of myxinoids. In: Brodal, A. & Fänge, R. (eds.) *The biology of Myxine*: Oslo: Universitetsforlaget; 516-549.
Hardisty, M.W. 1979. Biology of the cyclostomes. London: Chapman and Hall.
Hassan, A.Y. (1985) The interaction between drinking and feeding in young turbot, *Scophthalmus maximus* (L.) M.Sc. thesis, University of Wales.

Henderson, I.W. and Wales, N.A.M. 1974. Renal diuresis and antidiuresis after injections of arginine vasotocin in the freshwater eel (*Anguilla anguilla* L.). J. Endocr. 61, 487-500.

Hentschel, H. and Elger, M. 1987. The distal nephron in the kidney of fishes. Adv. Anat. Embryol. Cell. Biol. 108: 1-151.

Hentschel, H. (1987). Renal architecture of the dogfish *Scyliorhinus canicula* (Chondrichtthyes, Elasmobranchii). Zoomorphology 107: 115-125.

Larsen, L.O. and Dufour, S. 1993. Growth, reproduction and death in lampreys and eels. in Rankin, J.C. and Jensen, F.B. (eds.) *Fish Ecophysiology*: London: Chapman and Hall, 72-104.

Logan, A.G. 1978. Renal adaptation to salinity change in the river lamprey - a micropuncture study. Ph.D. thesis, University of Wales.

Logan, A.G., Moriarty, R.J., Morris, R. and Rankin, J.C. 1980. The anatomy and blood system of the kidney in the river lamprey, *Lampetra fluviatilis*. Anat. Embryol. 158: 245-252.

Logan, A.G., Moriarty, R.J. and Rankin, J.C. 1980. A micropuncture study of kidney function in the river lamprey, *Lampetra fluviatilis,* adapted to fresh water. J. exp. Biol. 85: 137-147.

Logan, A.G., Morris, R. and Rankin, J.C. 1980. A micropuncture study of kidney function in the river lamprey *Lampetra fluviatilis* adapted to sea water. J. exp. Biol. 88: 239-247.

McVicar, A.J. 1982. Renal and circulatory responses of the river lamprey (*Lampetra fluviatilis, L.)* to changes in environmental salinity. Ph.D. thesis, University of Wales.

McVicar, A.J. and Rankin, J.C. 1983. Renal function in unanaesthetised river lampreys (*Lampetra fluviatilis* L.): effects of anaesthesia, temperature and environmental salinity. J. exp. Biol. 105: 351-362.

McVicar, A.J. and Rankin, J.C. 1985. Dynamics of glomerular filtration in the river lamprey, *Lampetra fluviatilis* L. Am. J. Physiol. 249: F132-F138.

Moriarty, R.J. 1977. Renal function in the euryhaline lamprey (*Lampetra fluviatilis* L.) Ph.D. thesis, University of Wales.

Moriarty, R.J., Logan, A.G. and Rankin, J.C. 1978. Measurement of single nephron filtration rate in the kidney of the river lamprey, *Lampetra fluviatilis* L. J. exp. Biol. 77: 57-69.

Morris, R. 1958. The mechanism of marine osmoregulation in the lampern *Lampetra fluviatilis* L.) and the causes of its breakdown during the spawning migration. J. exp. Biol. 35: 649-665

Morris, R. 1965. Studies on salt and water balance in *Myxine glutinosa* (L.). J. exp. Biol. 42: 359-371.

Morris, R. 1972. Osmoregulation. in Hardisty, M.W. and Potter, I.C. (eds.) *The biology of lampreys*: Vol. 2: London: Academic press; 193-239.

Morris, R. & Pickering, A.D. 1975. Ultrastructure of the presumed ion-transporting cells in the gills of ammocoete lampreys, *Lampetra fluviatilis* (L.) and *Lampetra planeri* (Bloch). Cell. Tiss. 163: 327-341.

Nishimura, H., Oguri, M., Ogawa, M., Sokabe, H. and Imai, M. 1970. Absence of renin in the kidneys of elasmobranchs and cyclostomes. Am.J.Physiol. 218: 911-915.

Perrot, M.N., Grierson, C.E., Hazon, N. and Balment, R.J. 1992. Drinking behaviour in sea water and fresh water teleosts, the role of the renin-angiotensin system. Fish Physiol. Biochem. 10: 161-168.

Pickering , A.D. 1973. The measurement of unidirectional sodium fluxes in the non-stressed ammocoete river lamprey, *Lampetra fluviatilis* L. Comp. Biochem. Physiol. 44A: 613-623.

Pickering, A.D. and Morris, R. 1970. Osmoregulation of *Lampetra fluviatilis* L. and *Petromyzon marinus* (Cyclostomata) in hyperosmotic solutions. J. exp. Biol. 53: 231-243.

Pickering, A.D. and Morris, R. 1976. Fine structure of the interplatelet area in the gills of the macropthalmia stage of the river lamprey, *Lampetra fluviatilis* (L.). Cell. Tiss. Res. 168: 433-443.

Potts, W.T.W. and Parry, G. 1964. Osmotic and ionic regulation in animals. Oxford: Pergammon Press.

Potts, W.T.W., Fletcher, C.R. and Hedges, A.J. 1991. The in vivo transepithelial potential in a marine teleost. J. comp. Physiol. B. 161: 393-400

Rall, D.P. and Burger, J.W. 1967. Some aspects of hepatic and renal excretion in Myxine. Am. J. Physiol. 212: 354-356.

Rankin, J.C., Henderson, I.W. and Brown, J.A. 1983. Osmoregulation and the control of kidney function. in Rankin, J.C., Pitcher, T.J. and Duggan, R. T. (eds.) *Control processes in fish physiology*: London: Croom Helm; 66-88.

Rankin, J.C., Logan, A.G. and Moriarty, R.J. 1980. Changes in kidney function in the river lamprey, *Lampetra fluviatilis* L., in response to changes in external salinity. in Lahlou, B. (ed.) *Epithelial transport in the lower vertebrates*: Cambridge: Cambridge University Press; 171-184.

Rankin, J.C., Griffiths, V., McVicar, A.J. and Gilham, I.D. 1985. Control of kidney function in the river lamprey, *Lampetra fluviatilis*. in Lofts, B. and Holmes, W.N. (eds.) *Current trends in comparative endocrinology*: Hong Kong: Kong University Press; 893-895.

Rankin, J.C., Wahlqvist, I. and Wallace, B. 1984. Antidiuretic actions of angiotensin II, catecholamines and neurohypophysial hormones in the *in situ* perfused rainbow trout kidney. Gen. comp. Endocr. 53: 442.

Riegel, J. 1978. Factors affecting glomerular functions in the Pacific hagfish, *Eptatretus stouti* (Lockington). J. exp. Biol. 73: 261-277.

Riegel, J. 1986. Hydrostatic pressures in glomeruli and renal vasculature of the hagfish, *Eptatretus stouti.* J. exp. Biol. 123: 359-371.

Sinnott, R.J.H. and Rankin, J.C. 1976. Renal responses to transfer of the rainbow trout to sea water: possible involvement of neurohypophysial hormones. Gen. comp. Endocr. 29: 283.

Smart, G.R., Knox, D., Harrison, J.G., Ralph, J.A., Richards, R.H. and Cowey, C.B. 1979. Nephrocalcinosis in rainbow trout *Salmo gairdneri* Richardson; the effect of exposure to elevated CO_2 concentrations. J. Fish Dis. 2: 279-289.

Stolte, H. and Schmidt-Nielsen, B. 1978. Comparative aspects of fluid and electrolyte regulation by the cyclostome, elasmobranch and lizard kidney. In C. Barker Jørgensen and E. Skadhauge (eds.) *Proc. Alfred Benzon Symp. XI*: Copenhagen: Munksgaard; 209-220.

Suzuki, M., Kubokawa, K., Nagasawa, H. and Urano, A. 1995. Sequence analysis of vasopressin cDNAs of the lamprey, *Lampetra japonica*, and the hagfish, *Eptatretus burgeri*: evolution of cyclostome vasotocin precursors. J. mol. Endocrinol. 14: 67-77.

Uchiyama, M. and Murakami, T. 1994. Effects of AVT and vascular antagonist on kidney function and smooth muscle contraction in the river lamprey, *Lampetra japonica*. Comp. Biochem. Physiol. 107A: 493-499.

Uchiyama, M., Saito, N., Shimada, K. and Murakami, T. 1994. Pituitary and plasma arginine vasotocin levels in the lamprey, *Lampetra japonica*. Comp. Biochem. Physiol. 107A: 23-26.

Warne, J.M., Rankin, J.C. and Balment, R.J. 1993. Radioimmunoassay of plasma arginine vasotocin (AVT) concentrations in hyper- and hypo-osmoregulating lamprey (*Lampetra fluviatilis*). J. Endocr. 137: P156.

Weihprecht, H., Lorenz, J.N., Briggs, J.P. and Schnermann, J. 1991. Vasoconstrictor effect of angiotensin and vasopressin in isolated rabbit afferent arterioles. Am. J. Physiol. 261: F273-F282.

Weisbart, M., Dickhoff, Gorbman, A. and Idler, D.R. 1980. The presence of steroids in the sera of the pacific hagfish *Eptatretus stouti*, and the sea lamprey, *Petromyzon marinus*. Gen. comp. Endocr. 41: 506-519.

Youson, J.H. 1981. The kidneys In: Hardisty, M.W. and Potter, I.C., (eds.) *The biology of lampreys*: Vol. 3, London: Academic Press; 191-261.

Youson, J.H. 1982a. The morphology of the kidney in young adult anadromous sea lampreys, *Petromyzon marinus* L., adapted to seawater. I. General morphology and fine structure of the renal corpuscle and the proximal segments. Can. J. Zool. 60: 2351-2366.

Youson, J.H. 1982b. The morphology of the kidney in young adult anadromous sea lampreys, *Petromyzon marinus* L., adapted to seawater. II. Distal and Collecting segments, the archinephric duct, and the intertubular tissue and blood vessels. Can. J. Zool. 60: 2367-2381.

Ion and Water Balance in Elasmobranch Fish

N. Hazon, M.L. Tierney, G. Anderson, S. Mackenzie, C. Cutler and G. Cramb

Gatty Marine Laboratory, School of Biological and Medical Sciences, University of St Andrews, St Andrews, Fife KY16 8LB, Scotland

5.1 Introduction

Marine elasmobranchs maintain their plasma slightly hyperosmotic to that of the surrounding environment primarily due to the retention of the nitrogenous compound urea, the toxic effects of which are counterbalanced by the presence of trimethylamine oxide (TMAO). Although plasma sodium (Na) and chloride (Cl) concentrations are generally higher than those found in marine teleosts they are nevertheless lower than that of seawater (SW) and the fish face a continuous influx of NaCl across semi-permeable membranes in particular the gills. Due to the plasma hyperosmolarity some influx of water will occur and urea will be lost to the environment along a concentration gradient. The lesser spotted dogfish, *Scyliorhinus canicula*, feeds mainly on invertebrates while the spiny dogfish, *Squalus acanthias*, tend to feed sporadically on Euphausiids and other fish (Livingston, 1987, Tanasichuk *et al.*, 1991). Salt loading is presumably greatest during feeding when the ionic content of the food together with SW imbibed during feeding adds to the basal salt influx. Typically *S. canicula* tend to gorge food taking in large quantities over a short period of time following an interval of reduced feeding activity, and therefore the salt load associated with feeding is be intermittent. *S. canicula* must therefore regulate salt and water balance in the face of acute sodium and chloride challenges associated with large changes in dietary intake on an intermittent basis. The kidneys possess the capacity to produce a urine that is hyposmotic to SW and are not a major site of NaCl excretion; renal retention may be more a important function. Elasmobranchs are equipped with a specialised salt regulating gland, the rectal gland, that produces a NaCl secretion that is isoosmotic to SW. However, the feeding strategies of elasmobranchs does lead to ingestion of some water and salts. The intestinal distension which occurs as a result of the intake of food and/or the resulting high salt loading may act as a switch for the rectal gland which is known to be act intermittently.

Elasmobranchs also regulate their plasma osmolality in relation to changes in environmental salinity, but still maintain the plasma slightly hyperosmotic to the environment by varying plasma concentrations of urea and electrolytes. Upon gradual transferral to 50% SW the little skate, *Raja erinacea*, and the lip shark, *Hemiscyllium plagiosum*, show a decrease in plasma urea, which was attributed to a reduction in urea biosynthesis (Forster & Goldstein, 1976; Wong & Chan, 1977). A detailed study on *S. canicula*, was carried out by Hazon and Henderson (1984) where plasma urea, Na and Cl levels were all shown to decrease following gradual transferral of *S. canicula* to reduced salinity. Plasma urea reduction was attributed to a decrease in blood production rate of urea and a concomitant increase in plasma clearance rates of urea. Interestingly an increase in plasma urea, attributed to a decrease in plasma clearance rates, was observed following transferral of *S. canicula* to 140 % seawater (Hazon & Henderson, 1984). These studies suggest that regulation of plasma urea is a critical factor in the overall homeostasis of elasmobranch plasma. Using dietary protein restriction to constrain the urea-synthetic ability of *S. canicula*, Armour *et al.* (1993b) demonstrated that the normal osmoregulatory strategy was impaired in dogfish in 130% SW fed a low protein diet. These fish had no apparent ability to increase plasma urea concentration and instead adopted an alternative strategy of retaining increased plasma concentration. Clearly, in order to osmoregulate in the face of either changes in environmental salinity or dietary salt load elasmobranchs must closely co-ordinate the activities of the gill, kidney, gut and rectal gland. The osmoregulatory role and hormonal control mechanisms of these tissues will now be discussed.

5.2 Gills

The anatomy of elasmobranch gills has been studied by several authors (Wright, 1973; Olson and Kent, 1980; DeVries and DeJaeger, 1984; Metcalf and Butler, 1986). There are usually five pairs of gills with each gill arch consisting of a sheet of muscular and connective tissue (interbranchial septum) supported by lateral rods of cartilage (gill filament). Dorsal and ventral surfaces of each gill filament have a row of (secondary) lamellae and these are the principal site of gas exchange. Arterioarterial vasculature, which supplies the secondary epithelium, is controlled by sphincters located on efferent primary artery and also on both afferent and efferent arteries (Laurent and Dunel, 1980).

Chloride cells have been identified in elasmobranchs and described as a pear-shaped cell characterised by smooth endoplasmic reticulum, many mitochondria, and basolateral plasma membrane infoldings (Wright, 1973). More recently comparative studies of *Raja clavata* and *S. canicula* have revealed two types of chloride cells. In the former the chloride cell apical membrane is deeply buried in a *cul-de-sac* and connects to the external milieu by a narrow opening; in the second type the chloride cell apical membrane protrudes outward. However, the main characteristic common to both types is lack of a tubular system as found in teleots, which is functionally replaced by numerous basolateral membrane infoldings. Numerous mitochondria are peripherally disposed in close relationships with the infoldings of the basolateral cell membrane (Laurent and Dunel, 1980). Despite the presence of chloride cells however, elasmobranch branchial Na^{+}-K^{+}-ATPase activity was reported as being 10-15 times below that found in teleosts (Jampol & Epstein, 1970). A specific role for chloride cells in elasmobranch branchial epithelium has yet to be determined (Laurent & Dunel, 1980).

The permeability coefficient for urea in elasmobranch gill epithelia has been reported as one of the lowest in the animal kingdom (Boylan, 1967) however the precise mechanism have not been studied. To date some sort of "physical barrier" is considered to minimise urea loss across gill epithelia. Nonetheless, despite such an apparently efficient barrier quantative urea loss is still greatest across gill tissue (Shuttleworth, 1988), due to the large concentration gradient between plasma and sea water (Smith, 1936).

The elasmobranch gill epithelium has been much less studied with respect to movement of sodium and chloride ions than teleost gill epithelia. There is however a large inward concentration gradient for sodium and chloride ions between sea water and plasma. The absence of a substantial electropotential gradient across gill epithelia (Bentley *et al.*, 1976) led researchers to postulate that elasmobranchs were subject to a constant diffusional uptake of sodium and chloride ions across the gill epithelia (Maetz & Lahlou, 1966; Payan & Maetz, 1973).

Bentley *et al.* (1976) reported active sodium influx across the elasmobranch gill epithelia; while this may seem paradoxical it may be related to acid base balance. Ammonia is constantly being lost through the gill epithelium so for the elasmobranch to actively extrude this product required for urea production may appear inappropriate. However, the lack of an efficient buffer system and the capacity to restoring acid-base balance quickly could outweigh the loss of nitrogen through the sodium/ammonium (Na^+/NH_4^+) exchanger in the face of for example, severe hypercapnia (Caliborne & Evans, 1992). Bentley *et al.* (1976) reported that chloride ion influx was passive. However, regulation of chloride efflux has been demonstrated through a chloride/bicarbonate (Cl^-/HCO_3^-) exchange in elasmobranch gill epithelium where chloride efflux is coupled to bicarbonate influx during periods of hypercapnia (Randall *et al.*, 1976).

5.2.1 Hormonal control

There is very little information available with regards to hormonal control of ion transport in elasmobranch gills. In *S. acanthias* and *S. canicula* the responses of branchial vasculature to both adrenaline and noradrenaline appears to be primarily via a β-adrenoceptor-mediated vasodilation (Davies and Rankin, 1973; Capra and Satchell). This vasodilatory response masks a smaller α-adrenoceptor-meidated vasoconstriction (Davies and Rankin, 1973; Capra and Satchell, 1977).

5.3 Kidney

The kidneys in *S. canicula* and *S. acanthias* are found embedded in the dorsal side of the abdominal cavity as paired elongate organs either side of the dorsal aorta, which eventually fuse below the cloaca Elasmobranchs possess a renal portal system in which portal veins are formed from the bifurication of the large caudal vein, and on entry into the kidney divide and anastomose to form a matrix of small vessels. Portal blood mixes freely with blood from the glomerular vasa efferentia before leaving the kidney via the renal veins.

The functional element of elasmobranch renal tissue, the nephron, is a very long and complex tubular system and has been extensively reviewed in Lacy and Reale (1995). Ultrastructural analysis of the nephron in little skate *R. erinacea* (Lacy & Reale, 1991 a & b), and *S. canicula* (Hentschel *et al.*, 1993) provides strong evidence for a countercurrent system involving highly specialised and diverse epithelial transport.

There are two distinct regions of renal tissue in *R. erinacea* and *S. canicula*. 1) A dorsal "bundle" region that is enveloped by a urea impermeable sheath (Fig. 5.1) where it is thought that countercurrent exchange occurs (Stolte *et al.*, 1977); and 2) a ventral "sinus" region which exits from the bundle sheath to form two more distinct loops and a further possible site for countercurrent exchange (Fig. 5.1) (Lacy & Reale, 1991b). The nephron begins at the urinary pole of the renal corpuscle with a neck segment followed by proximal, intermediate, and distal segments and finally a collecting duct (Fig. 5.1). Each of these segments is further divided into subdivisions based on epithelial morphological differences. Epithelial differences within these groups include, presence of flagella, tight or gap junctions, brush borders, mitochondrial density, basolateral invagination and tubular dimension and rigidity. Lacy and Reale (1985 a & b, 1991 a & b) carried out detailed structural analysis of the nephron in the little skate, *R. erinacea*, where 16 morphologically different segments were assigned to the nephron tubule. However for the purpose of this discussion reference shall be made to the less complex model of the *S. canicula* nephron produced by Hentschel *et al.* (1993).

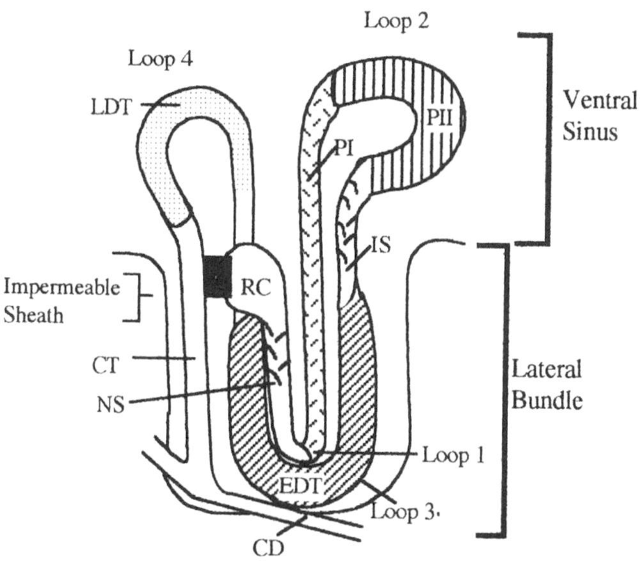

Fig. 5.1 Schematic diagram of a single nephron from *S. canicula*. Abbreviations: RC = renal corpuscle; NS = neck segment; PI and PII = proximal segments I and II respectivelly; IS - intermediate segment; EDT = early distal segment; LDT = late distal segment; CT = collecting tubule; CD = collecting duct.. Adapted from Hentschel *et al.* (1993). Anatomical Record, 235, 511-532. Reprinted by permission of Wiley-Liss, Inc., a subsidary of John Wiley & Sons, Inc.

Elasmobranchs produce a urine that is typically hyposmotic to blood plasma. The heterogeneity of tubular epithelial cells may be an adaptation for the retention of urea. Metabolic inhibitors such as phloretin and chromate have been shown to inhibit urea reabsorption, indicating active urea transport in the elasmobranch nephron (Hays *et al.*, 1977). Boylan (1967) demonstrated that only 35% of thiourea (a urea analogue) was reabsorbed in the elasmobranch nephron indicating extreme specificity for urea reabsorption. This led Forster (1970) to suggest an active urea reabsorption mechanism, and micropuncture studies have implicated the second proximal segment (PII) (Fig. 5.1) as a possible site of sodium-linked urea reabsorption (Stolte *et al.*, 1977). However a tentative model for passive urea reabsorption has also been proposed (Freidman & Hebert, 1990; Hebert & Freidman, 1990). This consisted of a proximal segment with a high water permeablity, a distal segment with active sodium transport but impereable to water and urea, a ventral loop with high water permeability and low urea permeability to allow water movement into the interstitium and a terminal tubule segment in the dorsal zone with a high urea permeability to allow passive diffusion of urea into the interstitium. Detailed analysis of urea permeabilities, and sites of transport along the elasmobranch nephron, is therefore required to ascertain whether urea reabsorption is either active, passive or both.

Urea transport in the elasmobranch kidney may certainly assist in regulation of urea plasma concentrations but the kidney also plays a role in the regulation of sodium and chloride plasma ion concentration. Ultrastructural studies have demonstrated that tubular cells in the early distal tubule (EDT) (Fig. 5.1) have characteristics very similar to cells that are known to actively transport sodium (Lacy & Reale, 1991b). It has been shown that loop 3 in the EDT exhibits similar characteristics of active sodium and chloride absorption associated with a renal diluting segment in mammals and amphibians (Freidman and Hebert, 1990; Hebert and Freidman, 1990). Reversible inhibition of sodium and chloride absorption with loop diuretics such as furosemide, low trans-epithelial resistance ("leakiness"), and negligible water permeability have also been shown (Fig. 5.1). In addition the presence of an ouabain sensitive Na+K+ATPase system was demonstrated (Hebert & Freidman, 1990).

Active sodium reabsorption has also been demonstrated in elasmobranch nephrons (Stolte *et al.*, 1977) utilising renal micropuncture, which implicated loop 2 in the second proximal region (PII) (Fig. 5.1) of *R. erinacea* to be the site of both sodium and chloride reabsorption (Stolte *et al.*, 1977). Furthermore isolated perfused PII tubules from *S. acanthias* have also been shown to actively secrete sodium and chloride ions (Beyenbach & Fromter, 1985) and this appears to drive net fluid secretion (Sawyer *et al.*, 1985 a & b). The ultrastructural evidence for active tubular reabsorption of sodium and chloride ions in the PII segment (Lacy *et al.*, 1975; Endo 1984) can also imply active tubular secretion of sodium and chloride ions (Henderson *et al.*, 1988). However histochemical studies have only isolated Na+K+ATPase in the early and late distal tubules (Fig. 5.1) and the collecting duct (Endo 1984; Hebert & Freidman 1990).

5.3.1 Hormonal control

Early studies, using primarily pharmacological doses of catecholamines, presented conflicting results on elamobranchs G.F.R. (Deetjen and Boylan, 1968; Forster *et al.*, 1972) but Brown and Green (1987) demonstrated a clear glomerular diuresis caused by adrenaline. The neurohypophysial hormone arginine vasotocin (AVT) has been identified in all species of elasmobranch investigated to date although little is known about its' physiological functions . However, the ability of elasmobranchs to alter renal tubular water permeability (Henderson *et al.*, 1988) suggests that an antidiuretic-like factor may be present.

5.4 Gut

The elasmobranch gut is densely innervated and many neuropeptides involved in controlling gut motility in mammals have been identified along the intestinal tract of the spiny dogfish *S. acanthias* (Holmgren & Nilsson, 1983). There is, however, no reported evidence in the literature of the elasmobranch gut playing an osmoregulatory role. This is perhaps due to the fact that elasmobranchs were previously thought not to drink (Smith 1936). However, using a sensitive method to determine drinking, Hazon *et al.* (1989) demonstrated drinking in elasmobranchs and pharmacological manipulation of the endogenous renin angiotensin system resulted in an increase in drinking (Hazon *et al.*, 1989).

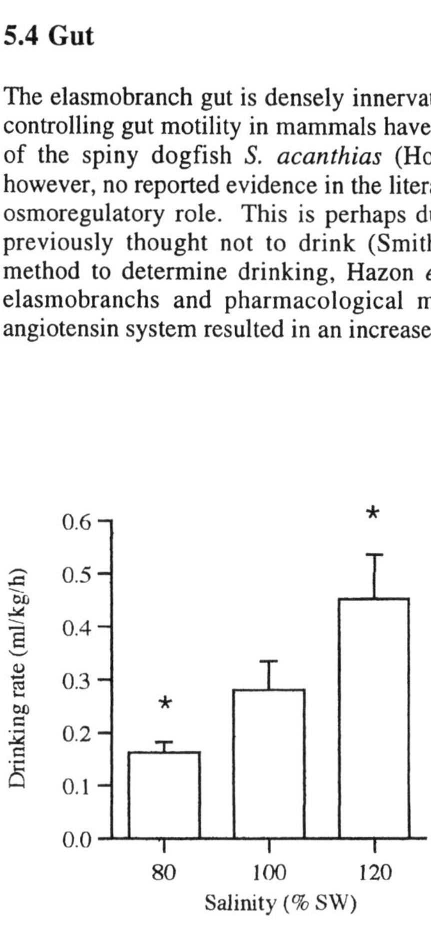

Fig. 5.2 Drinking rates in *S. canicula* adapted step-wise to varying environmental salinities. Results are the means ± S.E.M. of 10 animals. * indicates statistically significant differences at $p<0.01$ compared to 100% SW (unpaired t-test)

Furthermore using the same technique drinking rate of *S. canicula* have been shown to vary in response to changes in environmental salinity to which the animals are adapted (Fig. 5.2). To investigate the absorption of SW a NaCl load (1 ml SW), using^{22}Na (2µCi/ml) as marker molecule, was introduced into the intestine of *S. canicula* by means of a cannulae inserted into the intestine and the appearance of radioactivity in the blood plasma monitored. Sodium entered the bloodstream within one hour after bolus injection into the intestine. In view of this, it is therefore possible that the gut in elasmobranchs plays a more significant role in osmoregulation than first thought.

5.5 Rectal gland

The rectal gland, a small digitiform gland situated at the caudal end of the peritoneal cavity, secretes a fluid isosmotic to blood plasma but almost entirely composed of sodium chloride (Burger & Hess, 1960. Figure 5.3 illustrates the complex vascularization of the rectal gland. The gland is supplied by a single vessel, the rectal gland artery, which arises from the dorsal aorta. Vascular drainage from the gland is via a central venous sinus which, upon exiting the gland, follows the intestine anteriorally. The central duct of the gland opens into the lower intestine, posterior to the spiral valve and anterior to the rectum. The exiting duct and vein are closely linked, separated only by connective tissue and vessel walls

The structure of the rectal gland in sharks has been divided into 4 regions: 1) the outer fibro-muscular or capsular area with the principle arterial and venous systems: 2) the narrow sub-capsular layer consisting of smaller arterial and venous vessels, closely connected through arterio-venous anastomoses: 3) a middle glandular layer consisting of radially arranged tubules; and 4) a central region of longitudinally organised ducts arranged around a central canal and venous sinus (Bonting 1966; Kent & Olson, 1982; Masini *et al.*, 1994).

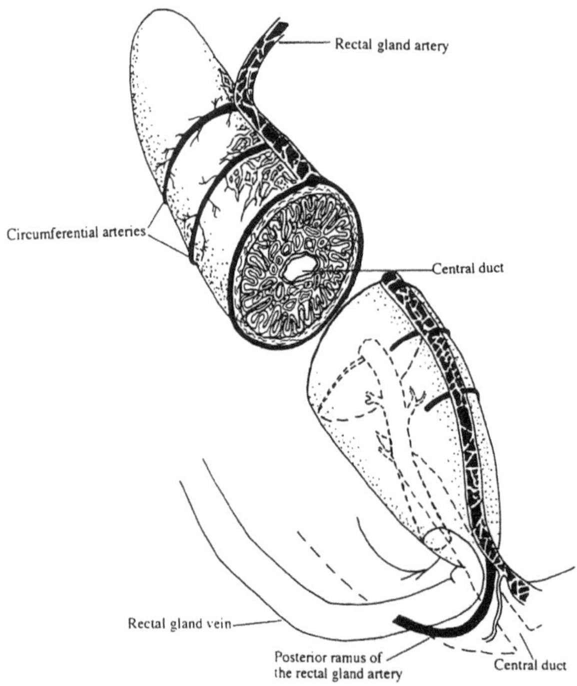

Fig. 5.3 Schematic diagram of the vascularization of the rectal gland. Adapted from Kent and Olson (1982), by kind permission of American Physiological Society, Bethesda, MD. Redrawen by Dr T.A. Beddow

Present evidence suggests that the net secretion of sodium chloride by rectal gland secretory epithelia is brought about through the combined action of four transport pathways, which together mediate secondary active transcellular chloride transport accompanied by paracellular sodium movement. Figure 5.4 illustrates the currently accepted pathways for sodium chloride secretion in rectal gland epithelia.

1) The basolateral Na^+-K^+-ATPase pump which creates the inwardly directed sodium gradient. Na^+-K^+-ATPase is found in high concentration on the basolateral membrane of rectal gland epithelia where the mitochondria also appears to be found (Goertemiller & Ellis 1976; Eveloff *et al.*, 1979). Silva *et al.* (1983) reported that stimulation of the isolated perfused rectal gland of *S. acanthias* significantly altered ouabain binding characteristics. Two classes of ouabain binding sites were suggested; one of high and one of low affinity. Stimulation appeared to increase the affinity of the high affinity binding sites, suggesting there was an immediate change in Na^+-K^+-ATPase configuration/activity/number upon stimulation, that regulated the overall secretory mechanism (Silva *et al.*, 1983).

2) The basolateral Na-K-2Cl co-transporter which utilises the sodium gradient to drive chloride ions above their electrochemical equilibrium. Passage of chloride ions into the lumen of the duct was in fact secondary to the action of a Na-K-2Cl co-transporter (Hannafin *et al.*, 1983). Furthermore the presence of a Na-K-2Cl co-transporter in the rectal gland tubule of *S. acanthias* was demonstrated using a series of inhibitors which illustrated that the Na-K-2Cl co-transporter was the only avenue for controlled sodium entry into the cell (Greger & Schlatter, 1984a & b). Immunofluorescent studies have also shown the Na-K-2Cl co-transporter to be distributed almost exclusively along the basolateral membrane and excluded from the apical side of the cell (Lytle *et al.*, 1992).

3) Potassium channels in the basolateral membrane which mediate potassium recycling. An ion channel, discovered on the basolateral membrane of the rectal gland epithelia (Greger and Schlatter, 1984a) was shown to extrude potassium into the extracellular fluid through the use of patch clamp techniques, thus providing a reservoir for the continuous recirculation of potassium through the cell, via the Na^+-K^+-ATPase pump and the Na-K-2Cl co-transporter (Greger *et al.*, 1987a). Known secretagogues were found to increase the binding of the radio-labelled loop-diuretic [^{3}H] benzmetanide to the Na-K-2Cl co-transporter in the perfused rectal gland and isolated tubules of *S. acanthias* (Forbrush *et al.*, 1992) an increase which was subsequently blocked by specific potassium channel blockers such as Barium (Ba^+).

4) Chloride channels in the apical membrane which permit passive loss of chloride ions into duct lumen. Patch clamping also identified two chloride channels on the apical membrane of the rectal gland cell, a large conductance channel considered only to be open when the rectal gland was stimulated (Greger *et al.*, 1987b) and a small conductance channel perhaps accounting for low chloride conductance in the unstimulated rectal gland (Gogelein *et al.*, 1987). A third chloride channel with close homology to human Cystic fibrosis transmembrane conductance regulator (CFTR) was also identified and localised in the elasmobranch rectal gland of *S. acanthias* (Marshall *et al.*, 1991). Investigation using sub-cellular fractions found this third, and smallest chloride channel, to be present in the highest frequency in epithelial cells of stimulated tubules isolated from the rectal gland of *S. acanthias* (Hanrahan *et al.*, 1993).

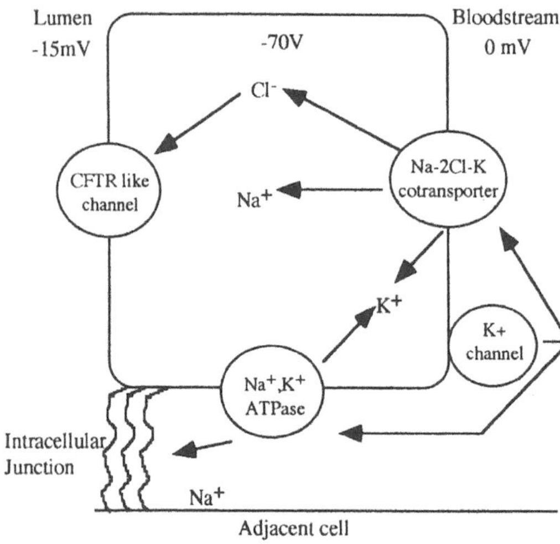

Figure 5.4 Schematic diagram showing the currently accepted pathways and membrane proteins involved in the active secretion of NaCl from rectal gland epithelium.

5.5.1 Hormonal control

One aspect of the intermittent behaviour of rectal gland activity is rapid activation of secretion, leading one to suggest the influence of neuroendocrine and/or autocrine factor(s) on rectal gland secretion. Vasoactive intestinal peptide-like (VIP) and bombesin-like (neurotransmitter) immunoreactive nerve fibres have been observed, following the radially arranged secretory tubules and also around the central excretory duct of the rectal gland of S. acanthias (Holmgren & Nilsson, 1983). Furthermore, VIP-like immunoreactivity has been found in the secretory parenchyma of the rectal gland of *S. acanthias* and *S. canicula* (Masini *et al*, 1994) indicating that VIP, as a neuropeptide, could have a direct effect on rectal gland secretion (Holmgren & Nilsson, 1983). VIP was reported to stimulate the accumulation of the second messenger cyclic adenosine monophosphate (cAMP) in isolated rectal gland cells of *S. acanthias* (Silva *et al.*, 1985). Although cAMP has been implicated in the stimulation of rectal gland secretionin elasmobranchs in general the mode of action is unclear (Stoff *et al.*, 1977; Silva *et al.*, 1977; Shuttleworth and Thompson 1980a and b; Lear *et al.*, 1992; Greger *et al.*, 1984; Hanrahan *et al.*, 1993). Reports to date on the regulation of rectal gland secretion in *S. canicula* and *S. acanthias* illustrate a major difference in the nature of gut peptides modulating rectal gland activity of these two elasmobranch species. Although there is little doubt that VIP induces an increase in secretion of the isolated perfused rectal gland and rectal gland tubules of *S. acanthias*, this has not been reported for other elasmobranchs. Neither heterologous nor homologous VIP stimulated rectal gland preparations *in vitro* in either *S. canicula* or *R. clavata* (Shuttleworth and Thorndyke 1984; Dimaline & Thorndyke, 1985; Thorndyke & Shuttleworth, 1985). Shuttleworth and Thorndyke (1984) attempted to isolate and characterise a homologous gut peptide that regulated rectal gland secretion. A stimulatory fraction, named rectin, was isolated but never characterised. Anderson *et al.* (1995) subsequently isolated and characterised a homologous gut peptide from *S. canicula* that was a potent stimulator of *in vitro* rectal gland activity and was discovered to be the tachykinin Scyliorhinin II (Scy II). It was postulated that rectin was in fact Scy II and that this tachykinin plays a central role in the control of rectal gland secretion.

The role of natriuretic factors in rectal gland function have been extensively studied. Atriopeptin, a synthetic analogue of mammalian atrial natriuretic peptide (ANP) tested *in vitro* was found to significantly increase chloride secretion rate (Solomon et al., 1985). Similar results were obtained from the isolated perfused rectal gland of *S. acanthias* following perfusion of rat atrial natriuretic peptide and endogenous heart extracts (Silva *et al.*, 1985). In elasmobranch fish CNP is the principal circulating natriuretic peptide (Suzuki *et al.* 1992). The endogenous C-type natriuretic peptide was found to stimulate secretion rate 7 to 8 times above basal levels in the isolated perfused rectal gland of *S. acanthias* (Solomon *et al.*, 1992).

Recently two types of CNP receptors have been identified on plasma membrane preparations of the rectal gland of *S. acanthias*, one which is linked to a guanylate cyclase pathway (therefore considered to be an active receptor), and the other, a low molecular weight receptor with no known links to a second messenger system (therefore considered to be inactive, possibly a 'clearance receptor') (Gunning *et al.*, 1993).

The intermittent nature of rectal gland secretion suggests that the blood flow through the gland during periods of quiescence is minimal which was indeed demonstrated by Kent and Olson (1982). It is possible that *in vivo* the vasculature of the rectal gland is likely to be influenced by a combination of the action of vasoconstrictive peptides, such as angiotensin II, and vasodilatory peptides, such as CNP.

In vitro perfusion studies in *S. acanthias* have identified two neuropeptides involved in the inhibition of rectal gland secretion. Infusion of the hypothalamic peptide, somatostatin in the explanted preparation utilised by Solomon *et al.* (1984b) was found to inhibit cAMP and forskolin-induced stimulation (Silva *et al.*, 1985). However, somatostatin did not affect VIP binding but did inhibit VIP stimulated accumulation of cAMP, although the precise site of action remained unclear (Silva *et al.*, 1985). A further study on the action of somatostatin was carried out by Silva *et al.* (1990) in which bombesin was used to investigate neuronal control of somatostatin on the perfused rectal gland of *S. acanthias*. Addition to the perfusate of bombesin alone produced no effect, however, when bombesin was added to glands previously stimulated with VIP, a reversible inhibition in chloride secretion was observed, in addition to a 10 fold increase in the liberation of somatostatin into the venous effluent. It was therefore postulated that bombesin inhibition of VIP-stimulated chloride secretion was secondary to the release of somatostatin (Silva *et al*, 1990). Neuropeptide Y (NPY) is a highly conserved neuropeptide found throughout the elasmobranchs and produces vasoconstrictor effects in the skate coronary artery and dogfish gill artery (Bjenning, 1992). Synthetic porcine NPY has been shown to inhibit secretion from whole gland, isolated tubules, and isolated cellular preparations of *S. acanthias* rectal gland stimulated by VIP, forskolin and cAMP (Silva *et al.*, 1993).

Furthermore, NPY inhibition of rectal gland secretion in *S. acanthias* was thought to be mediated at least in part by intracellular calcium concentration (Silva *et al.*, 1991).

The elasmobranch interrenal homologue 1α hydroxycorticosterone (1αOH-B) has also been implicated in the control of rectal gland secretion in elasmobranchs. Holt and Idler (1975) studied the effect of interrenalectomy and steroid replacement on the output of rectal gland fluid of the skate, *Raja ocellata.* It was found that after interrenalectomy there was a decrease in volume, osmolality, and sodium and chloride concentrations of rectal gland fluid. More recent studies have shown that 1αOH-B appears to regulate plasma sodium levels (Armour *et al.*, 1993 a & b) which supports the role of 1αOH-B as a steroid controlling rectal gland secretion rate.

5.6 Freshwater and Euryhaline Elasmobranchs

Elasmobranchs are traditionally thought of as exclusivly marine animals but true FW and euryhaline elasmobrachs do exist. Recent data with regards to these groups is sparse and has been largely analysed in previous reviews (Evans, 1979; Shuttleworth, 1988). Members of the ray family Potamotrygonidae (*Potamotrygon*, *Elipesurus*, and *Disceus*) which are found particularly in the Amazon and Orinoco river systems as much as 4000-4500 km from the sea (Thornson *et al.*, 1967) are the only elasmobranchs that can be genuinely considered as being permanently resident in FW. Plasma electrolyte levels resembles those of FW teleosts with plasma urea concentration approximately 1 mM and TMAO levels undetectable (Evans, 1979). Nitrogenous waste is mainly excreted as ammonia and not urea (Goldstein and Forster, 1971). *Potamotrygon* has the ability to survive in salinities up to approximately half SW due to retention of Na and Cl and not urea as urea biosynthetic ability and retention did not change (Thorson, 1970; Griffith *et al.*, 1973; Gerst and Thorson, 1977; Bittner and Lang, 1980).

Some euryhaline elasmobranchs, such as the sawfish *Pristis perotteti* (Thorson, 1967, 1982), originally thought to be land-locked are now known to be able to move freely between their FW habitats and the sea. Euryhaline elasmobranchs resident in FW show relatively small reductions in plasma ion concentrations and still retain, although reduced, levels of urea in their body fluids. Euryhaline species normally inhabiting a marine environment also show decreased urea levels in dilute salinities. The reduced urea levels are a result of a decline in the rate of urea synthesis, which can be reversed on exposure to full strength SW (Goldstein and Forster, 1971a; Wong and Chan, 1977). Euryhaline species possess the relatively low permeability to ions and high water permeability typical of marine elasmobranchs (Carrier and Evans, 1973).

5.7 Acknowledgements

Financial support from Biotechnology and Biological Sciences Research Council, U.K. and The Wellcome Trust is acknowledged. The authors would like to thank Dr T.A. Beddow for redrawing Figure 5.3.

References

Anderson, W.G., Conlon, J.M. and Hazon, N. (1995). Characterization of the endogenous intestinal peptide that stimulates the rectal gland of *Scyliorhinus canicula*. Am. J. Physiol., 268,

Armour K.J., O'Toole, L.B. and Hazon, N. (1993a). Mechanisms of ACTH- and angiotensin II-stimulated 1α-hydroxycorticosterone secretion in the dogfish, *Scyliorhinus canicula*. J. Mol. Endocr. 19, 235-244.

Armour K.J., O'Toole, L.B. and Hazon, N. (1993b). The effect of dietary protein restriction on the secretory dynamics of 1α hydroxycorticosterone and urea in the dogfish *Scyliorhinus canicula*: a possible role for 1α hydroxycorticosterone in sodium retention. J. Endocr., 138, 275-282.

Bentley, P.J., Maetz, J. and Payan, T. (1976). A study of the unidirectional fluxes of Na and Cl across the gills of the dogfish *Scyliorhinus canicula* (Chondrichthyes). J. Exp. Biol., 64, 629-637.

Bittner, A. and Lang, S. (1980). Some aspects of osmoregulation of Amazonian freshwater stingrays (*Potamotrygon hystrix*) I. Serum osmolality, sodium and shloride content, water content. hematocrit and urea level. Comp. Biochem. Physiol., 67A, 9-13.

Bjenning, C. (1992). Neuropeptide Y and bombesin in the cardiovascular system of fish. PhD thesis, University of Goteborg, Sweden.

Bonting, S.L. (1966). Studies on sodium-potassium-activated adenosine-triphosphatase.XV. The rectal gland of the elasmobranchs. Comp. Biochem. Physiol., 17, 953-966.

Boylan, J.W. (1967). Gill permeabliity in elasmobranchs. In "Sharks, skates and rays" (eds. P.W. Gilbert, R.F. Matthewson and D.P. Rall). John Hopkins University Press, Baltimore, pp 197-206.

Brown, J.A. and Green, C. (1987). Single nephron function of the lesser spotted dogfish, *Scyliorhinus canicula*, and effects of adrenaline. J. Exp. Biol., 129, 265-278.

Burger, J.W. and Hess, W.N. (1960). Function of the rectal gland in the spiny dogfish. Science, 131, 670-671.

Capra, M.F. and Satchell, G.H. (1977). The adrenergic resposnes of isolated saline perfused prebranchial arteries and gills of the elasmobranch *Squalus acanthias*. gen. Pharmacol., 8, 67-71.

Carrier, J.C. and Evans, D.H. (1973). Ion and water turnover in the freshwater elasmobranch *Potamotrygon* sp. Comp. Biochem. Phsyiol., 45A, 667-670.

Claiborne, J.B. and Evans, D.H. (1992). Acid-base balance and ion transfers in the spiny dogfish (*Squlaus acanthias*) during hypercapnia: A role for ammonia excretion. J. Exp. Zool., 261, 9-17.

Davies, D.T. and Rankin, J.C. (1973). Adrenergic receptors and vascular responses to catecholamines of perfused dogfish gills. Comp. Gen. Pharmacol., 60, 830-840.

Deetjan, P. and Antkowiak, D. (1970). The nephron of the skate, *Raja erinacea*. Bull. Mt Desert Isl. Biol. Lab., 10, 5-7.

Deetjan, P. and Boylan, J.W. (1968). Linear velocity and flow rte of tubular fluid in surface nephrons of *Squalus acanthias*. Bull Mt. Desert Isl. Biol. Lab., 12, 28-29.

DeVries, R. and DeJaeger, S. (1984). The gill in the spiny dogfish, *Squalus acanthias*: respiratory and non-respiratory function. Am. J. Anat., 169, 1-29.
Doyle, W.L. (1962). Tubule cells of the rectal salt-gland of *Urolophus*. Am. J. Anat., 111, 223-238.
Endo, M. (1984). Histological and enzymatic studies on the renal tubules of some marine elasmobranchs. J. Morph., 182, 63-69.
Evans, D.H. (1979). Fish. In Comparative Physiology of Osmoregulation in Animals (ed Malouy, G.M.O.) Academic Press, Orlando, pp 305.
Eveloff, J., Karnaky, K.J., Silva, P., Epstein, F.H. and Kinter, W.B. (1979). Elasmobranch rectal gland cell autoradiographic localisation of (3H) ouabain-sensitive Na,K-ATPase in rectal gland of the dogfish *Squalus acanthias*. J. Cell. Biol., 83, 16-32.
Forbrush, B., Haas, M. and Lytle, C. (1992). Na-K-Cl cotransport in the shark rectal gland 1. Regulation in the intact perfused gland. Am. J. Physiol., 262, C1000-C1008.
Forster, R.P. (1970). Urea and the early history of renal clearance studies. In "Urea dn Kidney (eds B. Schmidt-Nielsen and D.W.S. Kerr), Excerpta Medica Foundation, Amsterdam, pp 227.
Forster, R.P., Goldstein, L. and Rosen, S.K. (1972). Intrarenal control of urea reabsorpion by renal tubules of the marine elasmobranch *Squalus acanthias*. Comp. Biochem. Physiol., 42A, 3-12.
Forster,R.P. and Goldstein, L. (1976). Intracellular osmoregulatory role of amino acids and urea in marine elasmobranchs. Am. J. Physiol., 230, 925-931.
Friedman, P.A. and Hebert, S.C. (1990). Diluting segment in kidney of dogfish shark I: Localisation and characterisation of chloride absorption. Am. J. Phsyiol., 258, R398-R408.
Gerst, J.W. and Thorson, T.B. (1977). Effects of saline acclimation on plasma electrolytes, urea excretion and hepatic urea biosynthesis in a freshwater stingray, *Potamotrygon* so. Garman, 1877. Comp. Biochem. Physiol., 56A, 87-93.
Goertemiller, C.C. and Ellis, R.A. (1976). Localization of ouabain-sensitive, potassium-dependent nitrophenyl phosphatase in the rectal gland of the spiny dogfish, *Squalus acanthias*. Cell Tiss. Res., 175, 101-112.
Gogelein, H., Schlatter, E. and Greger, R. (1987). The "small" conductance channel in the luminal membrane of the rectal gland of the dogfish (*Squalus acanthias*). Pflugers Arch., 409, 122-125.
Goldstein, L. and Forster, R.P. (1971). Osmoregulation and urea metabolism in the little skate *Raja erinacea*. Am. J. Physiol., 220, 742-746.
Greger, R. and Schlatter, E. (1984a). Mechanism of NaCl secretion in the rectal gland of spiny dogfish (Squalus acanthias). I Experiments in isolated in vitro perfused rectal gland tubules. Pflugers Arch., 402, 63-75.
Greger, R. and Schlatter, E. (1984b). Mechanism of NaCl secretion in the rectal gland tubules of spiny dogfish (*Squalus acanthias*). II Effects of inhibitors. Pflugers Arch., 402, 364-375.
Greger, R., Gogelein, H. and Schlatter, E.(1987a). Potassium channels in the basolateral membrane of the rectal gland of *Squalus acanthias*. Pflugers Arch., 409, 100-106.

Greger, R., Schlatter, E. and Gogelein, H. (1987b). Chloride channels in the luminal membrane of the rectal gland of the dogfish (*Squalus acanthias*). Properties of the "larger" conductance channel. Pflugers Arch., 409, 114-121.

Greger, R., Schlatter, E., Wang, F. and Forrest, J.N. (1984a). Mechanism of NaCl secretion in rectal gland tubules of spiny dogfish (*Squalus acanthias*) III Effects of stimulation of secretion by cyclic AMP. Pflugers Arch., 402, 376-384.

Griffith, P.W., Pang, P.K.T., Srivastava, A.K. and Pickford, G.E. (1973). Serum composition of freshwater stingrays (Potamotrygonidae) adpated to fresh and dilute seawter. Biol. Bull., 144, 304-320.

Gunning, M., Cuero, C., Solomon, R. and Silva, P. (1993). C-type natriuretic peptide receptors and signalling in rectal gland of *Squalus acanthias*. Am. J. Physiol., 264, F300-F305.

Hannafin, J., Kinne-Saffran, E., Friedman, D. and Kinne, R.(1983). Presence of a sodium potassium chloride co-transport system in the rectal gland of Sualus acanthias. J. Memb. Biol., 75, 73-84.

Hanrahan,, J.W., Duguay, F., Samson, S., ALon, N., Jensen, T., Riordan, J.R. and Grzelczak, Z. (1993). Low-conductance chloride activated by cyclic AMP in the rectal gland of the shark Squalus acanthias and in cells heterologously expressing the shark CFTR gene. Bull Mt. Desert Isl. biol. Lab., 32, 48-49.

Hays, R.M., Levine, S.D., Myers, J.D., Heinemann, H.O., Kaplan, A., Franki, N. and Berliner, H. (1977). Urea transport in the dogfish kidney. J. Exp. Zool., 199, 309-316.

Hazon, N. and Henderson, I.W. (1984). Secretory dynamics of 1 α-hydroxycorticosterone in the elasmobranch fish, *Scyliorhinus canicula*. J. Endocr., 103, 205-211.

Hazon, N., Balment, R.J., Perrott, M. and O'Toole, L. (1989). The renin-angiotensin system and vascular and dipsogenic regulation in elasmobranchs. Gen. Comp. Endocr., 74, 230-236.

Hebert, S.C. and Friedman, P.A. (1990). Diluting segment in kidney of dogfish shark II: Electrophysiology of apical membranes and cellular resistances. Am. J. Physiol., 258, R409-R417.

Henderson, I.W., O'Toole, L.B. and Hazon, N. (1988). Kidney function. In "Phsyiololgy of Elasmobranch Fishes" (ed T.J. Shuttleworth), Springer-Verlag, Berlin, pp 203-214.

Hentschel, H. Mähler, S. and Elger, M. (1993).Renal tubule of dogfish, *Scyliorhinus caniculus*: A comprehensive study of structure with emphasis on intramembrane particels and immunoreactivity for H^+-K^+-Adenosine triphosphatase. The , 235, 511-532.Anatomical Record

Holmgren, S. and Nilsson, S. (1983). Bombesin-, gastrin/CCK-, 5-hydroxytryptamine-, neurotensin-, somatostatin-, and VIP-like immunoreactivity and catecholamine fluoresence in the gut of the elasmobranch *Squalus acanthias*. Cell and Tiss. Res., 234, 595-618.

Holt, W.F. and Idler, D.R. (1975). Influence of the interrenal gland on the rectal gland of the skate. Comp. Biochem. Physiol., 50C, 111-119.

Jampol, L.M. and Epstein, F.M. (1970). Sodium-potassium-activated adenosintriphosphatase and osmotic regulation by fishes. Am. J. Physiol., 218, 607-611.

Kent , B.and Olson, K.R. (1982). Blood flow in the rectal gland of *Squalus acanthias*. Am. J. Physiol., 243, 296-303.
Lacy, E.R. and Reale, E. (1985a). The elasmobranch kidney I: Gross anatomy and general distribution of nephrons. Anat. Embryol., 173, 23-34.
Lacy, E.R. and Reale, E. (1985b). The elasmobranch kidney II: Sequence and structure of the nephrons. Anat. Embryol., 13, 163-186.
Lacy, E.R. and Reale, E. (1991a). Fine structure of the elasmobranch renal tubule: Neck and proximal segments of the little skate. Am. J. Anat., 190, 118-132.
Lacy, E.R. and Reale, E. (1991b). Fine structure of the elasmobranch rneal tubule: Intermediate, distal and collecting duct segments of the little skate. Am. J. Anat., 192, 478-479.
Lacy, E.R. and Reale, E. (1995). Functional morphology of the elasmobranch nephron and retention of urea. In Cellular and Molecular Approaches to Fish Ionic Regulation (eds C.M. Wood and T.J. Shuttleworth). Academic Press, New York, pp 107-146.
Lacy, E.R., Schmidt-Neilsen, B., Galaske, R.G. and Stolte, H. (1975). Configuration of the skate (Raja erinacea) nephron and ultrastructure of two segments of the proximal tubule. Bull. Mt. Desert Isl. biol. lab., 15, 54-56.
Laurent, P. and Dunel, S. (1980). Morphology of gill epithelia in fish. Am. J. Physiol., 238, R147-R159.
Lear, S., Cohen, B.J., Silva, P., Lechene, C. and Epstein, F.H. (1992). cAMP activates the sodium pump in the cultures cells of the elasmobranch rectal gland. J. Am. Soc. Nephrology, 2 (10), 1523-1528.
Livingston, M.E. (1987). Spawning hoki (*Macruronus novaezelandiae hector*) concentrations in Cook Strait and off the east-coast of the South Island, New Zealand, August - September 1987. New Zealand J Marine and Freshwater Research, 24 (4), 503-517.
Lytle, C., Xu, J.C., Biemesderfer, D., Haas, M. and Forbrush, B. (1992). The Na-K-Cl cotransport protein of shark rectal gland I. Development of monoclonal antibodies, immunoaffinity purification and partial biochemical characterisation. J. Biol. Chem., 267 (35), 25428-25437.
Maetz, J. and Lahlou, B. (1966). Les echanges de sodium et de chlore chez un elasmobranche *Scyliorhinus*, mesures a l'aide des isotopes ^{24}Na et ^{36}Cl. J. Physiol. (Paris), 58, 249.
Marshall, J., Martin, K.A., Picciotto, M., Hockfield, S., Nairn, A.C. and Kaczmarak, L.K. (1991). Identification and localization of a dogfish homolog of human cystic fibrosis transmembrane conductance regulator. J. Biol. Chem., 266 (33), 22749-22754.
Masini, M.B.A., Uva, B., Devecchi, M. and Napoli, L. (1994). Renin-like activity, angiotensin I converting enzyme-like activity, and osmoregulatory peptides in the dogfish rectal gland. Gen. Comp. Endocr., 93, 246-254.
Metcalf, J.D. and Butler, P.J. (1986). The functional anatomy of the gills of the dogfish (Scyliorhinus canicula). J. Zool., 208, 519-530.
Ng, K.K.F. and Vane, J.R. (1967). The conversion of angiotensin I to angiotensin II. Nature (London), 216, 762-766.
Olson, K.R. and Kent, B. (1980). The microvasculature of the elasmobranch gill. Cell Tiss. Res., 209, 49-66.

Payan, P. and Maetz, J. (1973). Branchial sodium transport mechanisms in Scyliorhinus canicula: evidence for Na^+/NH_4^+ and Na^+H^+ exchanges and for a role of carbonic anhydrase. J. Exp. Biol., 58, 487-502.
Randall, D.J., Heisler, N. and Drees, F. (1976). Ventilatory response to hypercapnia in the lesser spotted dogfish *Scyliorhinus stellaris*. Am. J. Phsyiol., 230, 590-594.
Sawyer, D.B., Cliff, W.H., Wilhelm, M.M., Fromter, R.O. and Beyenbach, K.W. (1985a). Proximal tubules of the glomeular shark kidney secrete fluid via secretion of NaCl. Fed. Proc., 44, 8688.
Sawyer, D.B., Cliff, W.H., Wilhelm, M.M., Fromter, R.O. and Beyenbach, K.W. (1985b). Mechanism of fluid secretion by proximal tubules in the glomeular kidney of the shark. Kidney Int., 27, 319.
Shuttleworth, T.J. (1983). Haemodynamic effects of secretory agents on the isolated elasmobranch rectal gland. J. Exp. Biol., 103, 193-204.
Shuttleworth, T.J. (1988). Salt and water balance. In "Physiology of Elasmobranch Fishes" (ed T.J. Shuttleworth), Springer-Verlag, London, pp 171-199.
Shuttleworth, T.J. and Thompson, J.L. (1980). Oxygen consumption in the rectal gland of the dogfish *Scyliorhinus canicula*, and the effects of cyclic AMP. J. Comp. Physiol., 136, 39-43.
Shuttleworth, T.J. and Thorndyke, M.C. (1984). An endogenous peptide stimulates secretory activity in the elasmobranch rectal gland. Science, 225, 319-321.
Silva, P., Epstein, F.H. and Solomon, R.J. (1993). C-type natriuretic peptide (CNP) stimulates chloride secretion by rectal gland of the shark via a dual mechanism of action involving protein kinase C. Clin. Res., 41 (2), 327A.
Silva, P., Epstein, J.A., Stevens, A., Spokes, K. and Epstein, F.H. (1983). Ouabain binding in the rectal gland of *Squalus acanthias*. J. Membr. Biol., 75, 105-114.
Silva, P., Lear, S., Reichlin, S. and Epstein, F.H. (1990). Somatostatin mediates bombesin inhibition of chloride secretion by rectal gland. Am. J. Physiol., 258, R1459-R1463.
Silva, P., Solomon, R.J., Landsberg, J., Hervieux, A., Emmonds, R. and Epstein, F.H. (1991). Effect of calcium channel blockers on inhibition of chloride secretion in the shark rectal gland. Bull. Mt. Desert Isl. biol. lab., 30, 67-70.
Silva, P., Stoff, J., Field, L., Fine, L., Forrest, J.N. and Epstein, F.H. (1977). Mechanism of active chloride secreiton by shark rectal gland: role of Na-K-ATPase in chloride transport. Am. J. Physiol., 233, F298-F306.
Silva, P., Stoff, J.S., Leone, D.R. and Epstein, F.H. (1985). Mode of action of somatostatin to inhibit secretion by shark rectal gland. Am. J. Physiol., 249, R329-R334.
Smith, H.W. (1936). The rentention and physiological role of urea in the Elasmobranchii. Biol. Rev., 11, 49-82.
Solomon, R., Protter, A., McEnroe, G., Porter, J.G. and Silva, P. (1992). C-type natriuretic peptides stimulate chlordie secretion in the rectal gland of Squalus acanthias. Am. J. Physiol., 262, R707-R711.

Solomon, R.J., Taylor, M., Rosa, R., Silva, P. and Epstein, F.H. (1984a). *In vivo* effect of volume expansion on rectal gland function. II Hemodynamic changes. Am. J. Physiol., 246, R61-R71.

Solomon, R.J., Taylor, M., Stoff, J.S., Silva, P. and Epstein, F.H. (1984b). *In vivo* effect of volume expansion on rectal gland function I. Humoral factors. Am. J. Physiol., 246, R63-R66.

Solomon, R., Taylor, M., Dorsey, D., Silva, P. and Epstein, F.H. (1985). Atriopeptin stimulaion of rectal gland function in *Squalus acanthias*. Am. J. Physiol., 249, R348-R354.

Stoff, J.S., Silva, P., Field, M., Forrest, J.N., Stevens, A. and Epstein, F.H. (1977). Cyclic AMP regulation of active chloride transport in the rectal gland of marine elasmobranhs. J. Exp. Zool., 199, 443-448.

Stolte, H., Galaske, R.G., Eisenbach, G.H., Lechene, C., Schmidt-Nielsen, B. and Boylan, J.W. (1977). Renal tubule ion transport and collecting duct function in the elasmobranch little skate *Raja erinacea*. J. Exp. Zool., 199, 403-410.

Suzuki, R, Togashi, K, Ando, K and Takei, Y (1992). Distribution and molecular forms of C-type natriuretic peptide in plasma and tissue of a dogfish, *Triakis scyllia*. Gen. Comp. Endocr., 96, 378-384.

Tanasichuk, R.W., Ware, D.M., Shaw, W. and McFarlane, G.A. (1991). Variations in diet, daily ration, and feeding periodicity of Pacific hake (*Merluccius productus*) and spiny dogfish (*Squalus acanthias*) off the lower west-coast of Vancouver Island. Canadian J. Fisheries and Aquatic Sciences, 48 (11), 2118-2128.

Thorndyke, M.C. and Shuttleworth, T.J. (1985). Biochemical and physiological studied on peptides from the elasmobranch gut. Peptides, 6(3), 369-372.

Thorson, T.B. (1967). Osmoregulation in fresh-water elasmobranchs. In Sharks, Skates and Rays. (eds P.W. Gilbert, R.F. Matheson and D.P. Rall. John Hopkins, Baltimore, pp 265-270.

Thorson, T.B. (1970). Freshwater stingrays, *Potamotrygon* spp: failure to concentrate urea when exposed to saline medium. Life Sci., 9, 893-900.

Thorson, T.B. (1982). Life history implications of a tagging study of the large tooth sawfish, *Prislis perottete*, in the Lake Nicaragua- Rio San Jan system. Environ. Biol. Fishes, 7, 207-228.

Thorson, T.B., Cowan, C.M. and Watson, D.E. (1967). *Potamotrygon* spp: Elasmobranchs with low urea content. Science 158, 375-377.

Wong, T.M. and Chan, D.K.O. (1997). Physiological adjustments to dilution of the external medium in the lip-shark Hemiscyllium plagiosum (Bennett) II: Branchial, renal and rectal gland function. J. Exp. Zool., 200, 85-96.

Wright, D.D. (1973). The structure of the gills of the elasmobranch *Scyliorhinus canicula*. Z. Zellforch., 144, 489-509.

Gill Chloride Cells Activation by Plasma Osmolarity

J.A.Zadunaisky

Rosenstiel School of Marine and Atmospheric Science, University of Miami, Miami, FL, 33149, and the Mount Desert Island Biological Laboratory, Sallsbury Cove, Me 04672 USA

6.1 Introduction and background

The fish gill is the most important organ for salt excretion and absorption in sea water fish, besides the all important respiratory function (Smith 1930). The interest in the specific cellular epithelium performing secretion started in the early 1930's when this function of the gill was tested in eels by perfusing a salt solution through the circulation and measuring the salt concentration in a closed circuit bath where the fish were placed (Keys, 1931). The results showed that salts, as measured by the increase in total chlorides, were excreted from the plasma to the bath in which the eels were placed. A special cell or cellular epithelium was assumed to exist because of the lack of structural elements, such as organelles, mitochondria, and an elaborate endoplasmic reticulum in the respiratory cells. In fact an histological search for cells with these secretory characteristics permitted the finding of mitochondria

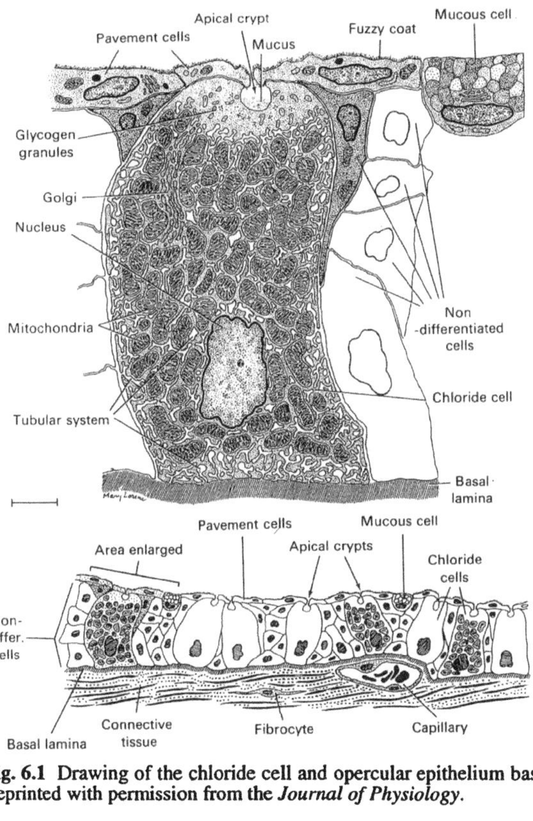

Fig. 6.1 Drawing of the chloride cell and opercular epithelium based on multiple electron micrographs. Reprinted with permission from the *Journal of Physiology*.

rich cells located at the base of the filaments of the gills. These cells not only were present in sea water fish, but in the case of the eel, they were atrophic in eels adapted to fresh water (Keys and Wilmer 1932). Due to the change in chloride concentration found in the perfusion experiments, the newly found cells were serendipitously called "chloride cells". In fact for many years they were refered to by the sceptical as "mitochondria rich" cells (see Bevelander 1936). We now know that they actually secrete sodium chloride, by means of an active transport mechanism that pumps chloride ions from blood to sea water (see Zadunaisky 1984 for review).

The anatomy of the chloride cell is shown in Fig. 6.1 . Salient features include an elaborate endoplasmic reticulum, the characteristic pit at the apical side of the cell, the great number of mitochondria in the cytoplasm and the elaborate indentations of the cell membranes in contact with other chloride cells of the epithelium at the base of the gill filaments. There are tight junctions facing the apical side of the epithelium and the cells rest on a basement membrane. This description of the cell presented in Fig. 6.1 corresponds to both chloride cells of the gill epithelium and what could be called aberrant or migrated chloride cells that are found on the epithelium of the roof of the mouth and on the opercular epithelium, lining the inside of the flap covering the gills (Burns and Copeland, 1950).

The opercular epithelium has been dissected out, mounted in Ussing type chambers and chloride secretion detected by measuring the unidirectional fluxes of radioactive chloride. All the electrical current produced by mounted operculii is carried by chloride ions with sodium following passively. This finding has been made in opercular preparations containing a high number of chloride cells, mainly from species that are euryhaline and survive both in sea water and fresh water (Degnan, Karnaky and Zadunaisky 1977; Marshall, 1977 and Foskett et al 1981). The chloride cells belong to a family of secretory cells all with similar if not identical characteristics. They are encountered in the distal tubules of the kidney, the salt secreting nasal gland of birds, the chloride secretory cells of the shark rectal gland, the gastric and intestinal chloride secretory cells, the cells of the choroid plexus of the brain and the ciliary epithelium of the eye, as well as the salivary and lachrymal glands. Chloride secretory cells of the corneal epithelium have mitochondria but, because of their location and the need for transparency, have a less elaborate endosome.

The rate of chloride secretion is a function of the requirements of the homeostatic mechanisms that control the constant composition of the internal medium. Chloride secretion increases when fish are exposed to high salinity and stops when placed in fresh water. The best experimental species are euryhaline because during the transitions from low to high salinity the system of secretion is activated. Fig. 6.2 shows the change in plasma osmolarity in *Fundulus heteroclitus,* when specimens of this nordatlantic fish are transferred from fresh water to sea water. There is a rapid increase in total osmolarity, reflecting an increase mostly in sodium chloride and after a peak, there is a decrease and adjustment to preexisting conditions. The adaptation is due to a faster rate of chloride secretion through the gill chloride cells. This returns the original plasma level of salts to approximately 300 mOsm.

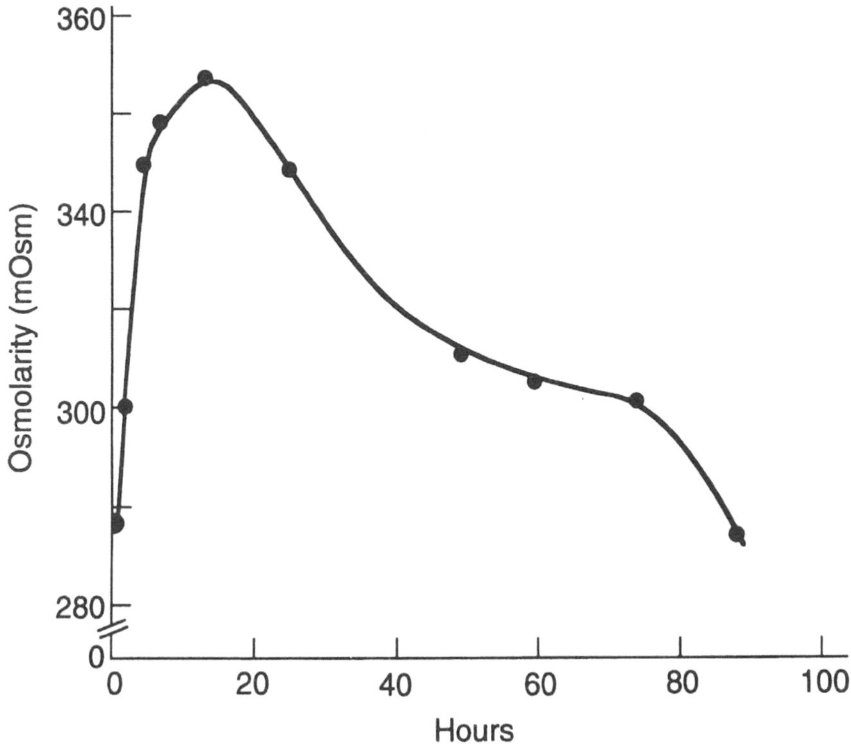

Fig. 6.2. Plasma osmolarity of *F. heteroclitus* transferred at 0 time form fresh water to sea water. Each point corresponds to the osmolarity of a plasma sample pooled from blood of 7 fish. Note the increase of 65 mOsm at the peak and the gradual return to normal plasma values. Data reprinted with permission from the *Journal of Membrane Biology*.

The reason for the increase in salts in the plasma is that the fish are required to swallow the sea water that is then absorbed in the intestine and increases plasma osmolarity. This rapid change in osmolarity triggers the chloride secretion of the chloride cells and it can be considered as the *rapid signal* the cells receive to start or increase their pumping rate. After several days have gone by there is another response that consists in the increase in the number of chloride cells in the gill epithelium and anywhere else that this cells are found. The *slow signal* is hormonal in nature and has been well documented that cortisone is responsible for the proliferation and the activation of the cells (Maetz and Bornicin 1975). Thus in experiments designed to exaggerate the rapid as well as the slow signals, *Fundulus heteroclitus* are transferred from fresh water to 2x sea water, that is an artificial sea water containing twice the total salts of natural sea water and therefore twice its osmolarity. The result is a dramatic proliferation in several days of the chloride cells of the gill epithelium that now not only appear at the base of the gill filaments but pervade the whole length of some filaments. In this process, the respiratory epithelium is covered with chloride cells and the fish cannot maintain a normal gas exchange and perish essentially of anoxic anoxia. Another change that occurs is the increase in the transport enzyme Na^+,K^+ATPase, that has been shown by measurements of ouabain binding sites to increase dramatically under these conditions (Karnaky et al 1976).

Until recently (Zadunaisky *et al* 1995) the nature of the rapid signal received by the chloride cells to start secreting more chloride during the transitions to high salinity or to reduce the rate of pumping when transferred from sea water to fresh water was not clear . As will be discussed here, the change in plasma osmolarity that occurs in these transitions is the rapid signal for adaptation. We will see that even small changes in total osmolarity on the basolateral or blood side of mounted opercular preparations produce dramatic changes in chloride secretion. In the mid 1970s our laboratory started a series of experiments with the chloride cell rich opercular epithelium of *Fundulus heteroclitus* mounted in small Ussing chambers. The mounting of a tiny epithelium on an aperture of less than 1 mm diameter was possible because we had developed special chips to mount another fragile and small epithelium, the retina pigment epithelium of frogs (Zadunaisky and Degnan, 1976). The work proceeded in collaboration with K. Karnaky and K. Degnan at the Mount Desert Island Biological Laboratory and at New York University in New York City. We demonstrated that the chloride cell rich epithelium when mounted between identical salt solutions produced a potential difference that was electrically negative on the sea water side and positive on the basolateral side. This voltage could be short-circuited and the current reached levels indicating net charge transfer of more than 4 μEq/sq.cm/hr which in biological transport is quite a large number. The charge transfer was studied utilizing radioisotopic fluxes of chloride, with Cl^{-36}, and of sodium with Na^{-22} (Degnan, Karnaky and Zadunaisky, 1977). The electrical current was carried only by the chloride ions, sodium moving passively across the tissue. An analysis of the paracellular pathway demonstrated that sodium was moving across this paracellular region and that the transfer numbers indicated about half of the total electrical current to be carried by sodium in open circuit conditions (Degnan and Zadunaisky, 1980 a,b). We also found that the epithelium required 16 mM of sodium bicarbonate for maximal current expression. It was not clear then but it is now, that this was due to the existence of a Cl^-/HCO_3^- exchanger in the basolateral side (Zadunaisky et al 1995). Chloride secretion was inhibited by addition of ouabain to block Na^+,K^+ATPase and by furosemide and bumetanide used to block the 2Cl,Na,K cotransporter. The preparation was stimulated by adrenergic hormones or their analogues. Thus epinephrine produced a biphasic response on the chloride current, excitation by activation of beta receptors and inhibition by alpha receptors. Cholinergic antagonists, acetylcholine, carbachol, and muscarine produced inhibition and the effect was shown to occur by activation of muscarinic receptors (Rowing and Zadunaisky 1978). The chloride cell membrane also has receptors for ANF, the natriuretic factor (Scheidi and Zadunaisky, 1988).

An experimental plan to examine the different transporting elements of the chloride cell was implemented and the findings indicate the following: 1) chloride secretion requires the presence of Na and K on the basolateral side, and together with 2) the inhibitory action of loop diuretics and bumetamide point to the presence of the 2Cl,Na,K cotransporter in the chloride cell membrane 3) ouabain binding localized the

Na^+,K^+ATPase to the basolateral side of the chloride cell 4) measurements of intracellular electrical potential difference and the resistance of the apical membrane during the action of isoproterenol (Zadunaisky, Curci, Sheidi and Schettino,1988) localize the chloride channels to the apical membrane of the chloride cell, findings that in conjunction with 5) the existence of chloride channels detected in the cell membranes by the patch clamp technique in cultures of opercular chloride cells (Marshall et al,1995) corroborate the existence of chloride channels in the apical membrane 6) the effects of complete removal of bicarbonate suggest that a Cl^-/HCO_3^- exchanger exists in the basolateral membranes 7) the action of the inhibitor DIDS, highly specific for the Cl^-/HCO_3^- exchanger on the basolateral side confirm the presence of this exchanger (Zadunaisky et al 1995) 8) the action of the inhibitor amiloride at high concentrations indicates the existence of a Na^+/H^+ exchanger in the basolateral membranes of the chloride cells and 9) the activation by several specific drugs(see below) confirm the presence of the Na^+/H^+ exchanger.

A picture then emerges of the chloride cell as a chloride secretory type of the variety discussed above. Together with the main receptors for hormones, the model shown in Fig. 6.3 describes the several membrane transporters, exchangers and receptors involved in the basic functions of the chloride cell.

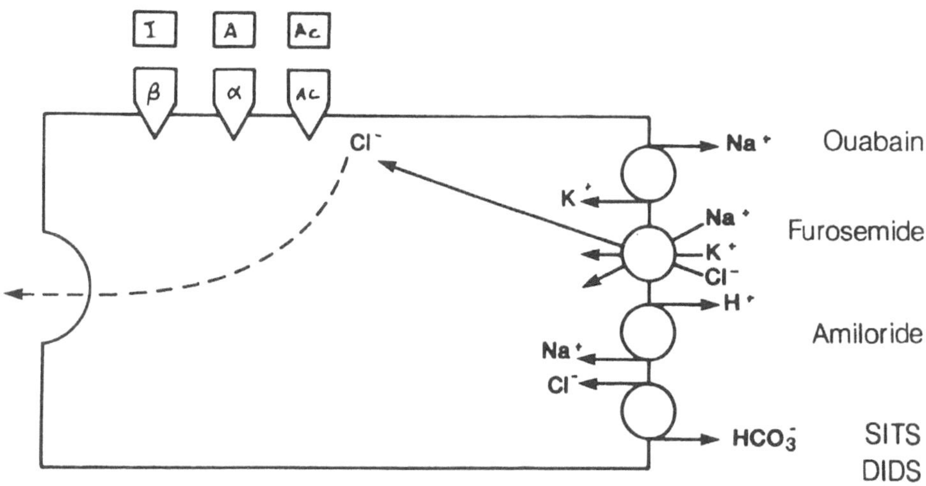

Fig. 6.3. Diagram of the chloride cell showing the membrane transport elements found so far. Reproduced with permission from the *Journal of Membrane Biology*.

6.2 The importance of osmolarity

What is the nature of the *rapid signal* for the activation or reduction in chloride secretory rate of the chloride cells?

In order to answer this question we conducted a search that indicates that the principal factor is the osmolarity of the plasma. The concentration of ions as well as osmolarity was examined in plasma of *Fundulus heteroclitus* during the transitions from fresh to sea water. The variations found were tested on both the apical and basolateral side of opercular epithelium preparations mounted in Ussing chambers where the chloride current was measured electrically. Changes in salinity or ion concentrations in the apical or sea water side did not produce changes in chloride secretion; instead, modifications of ion concentration or total osmolarity on the basolateral side produced the important effects.

6.2.1 Changes in plasma osmolarity during the transition to high salinity

The total osmolarity of the plasma of *Fundulus heteroclitus* adapted to fresh water and immediately after the transition to full sea water was measured by freezing point depression. Fig. 6.2 shows this curve described in the Introduction. The increase in osmolarity is of the order of 50 to 60 mOsm in plasma. This represents an increase of 25 to 30 mM of salts in the plasma, mostly NaCl at the peak of the transition. An increase of 50 mOsm was selected to be tested on the opercular epithelium. The results are discussed below.

6.2.2 Effects of the addition of 50 mOsm of mannitol on the opercular epithelium

In Fig. 6.4 it can be observed that the addition of 50 mOsm of mannitol to the basolateral side of an opercular preparation produces a dramatic increase in the chloride current that practically doubles the original resting value. It can also be seen that the addition of the same amount of mannitol to the apical side has no effects. The choice of mannitol, a non metabolizable, non electrolyte sugar rather than NaCl, is obvious because mannitol will not penetrate the cells and it will exert its full osmotic effect without interference with the actual process of ion secretion.

The results then imply that the increased osmolarity elicited by mannitol has produced shrinkage of the chloride cells. We have actually measured the volume changes by quantitative microscopy as explained further on in this paper. The changes induced on cells by hypertonic or hypotonic solutions have recently fallen under the subject of *cell volume regulation* (see review of Hoffman and Simonsen 1989). Cells tend to counteract the swelling or shrinkage produced by exposure to hypotonic or hypertonic solutions by returning to their original volume even if continuously exposed to the abnormal osmolarity. They accomplish this volume regulation by opening channels in the cell membrane for specific ions. For example in the most common case of cell volume increase after exposure to hypotonic solutions chloride and potassium channels are opened. The exit of the salt KCl produces the necessary

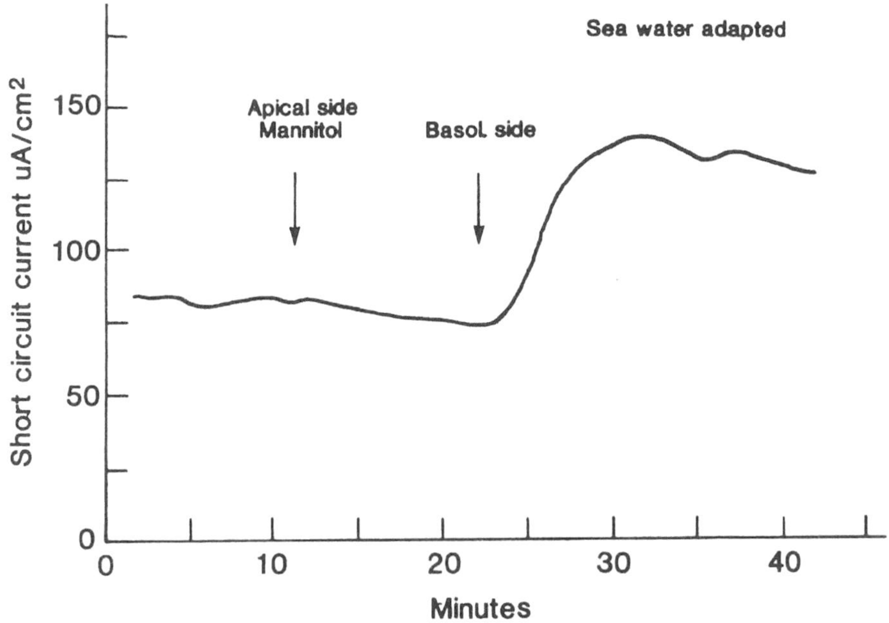

Fig. 6.4 *Increase of chloride secretion produced by 50 mOsm of mannitol in an isolated operculum of F. heteroclitus.* No effect is found on the apical side, but shrinkage of the chloride cells induced on the basolateral side produce rapid current stimulation. Data reprinted with permission for the *Journal of Membrane Biology.*

produces the necessary osmotic gradient to make water move out of the cells and they return to their initial volume. The result is a cell with reduced intracellular activity of KCl and normal volume. In the case of the exposure to hypertonic solutions the most common reaction is the activation of the 2Cl,K,Na cotransporter that brings salt into the cells as well as water with the return to normal volume in a process known as regulatory volume increase. In this process intracellular pH also plays an important role. The Na/H exchanger is silent in normal cell volume and pH , but is activated by changes in tonicity and, together with the Cl^-/HCO_3^- exchanger, tends to control cell pH and also cell volume.

The cases described above apply mostly to single cells such as red cells, or suspensions of ascitis tumor cells but in the case of epithelia similar rules apply. However in the chloride cell epithelium of the operculum the activation or inhibition of membrane transporters, exchanger or cotransporters, produces dramatic changes in the secretion of chloride as measured by the increases and decreases of the short circuit current.

This chloride current in the chloride cells of the opercular epithelium reflects intracellular changes of pH, as much as the sodium current of the frog skin epithelium reflects similar pH changes (Harvey, Thomas and Ehrenfeld 1988). Intracellular pH can control Na and K permeability of model epithelia, and in the case of the chloride cell it controls chloride permeability.

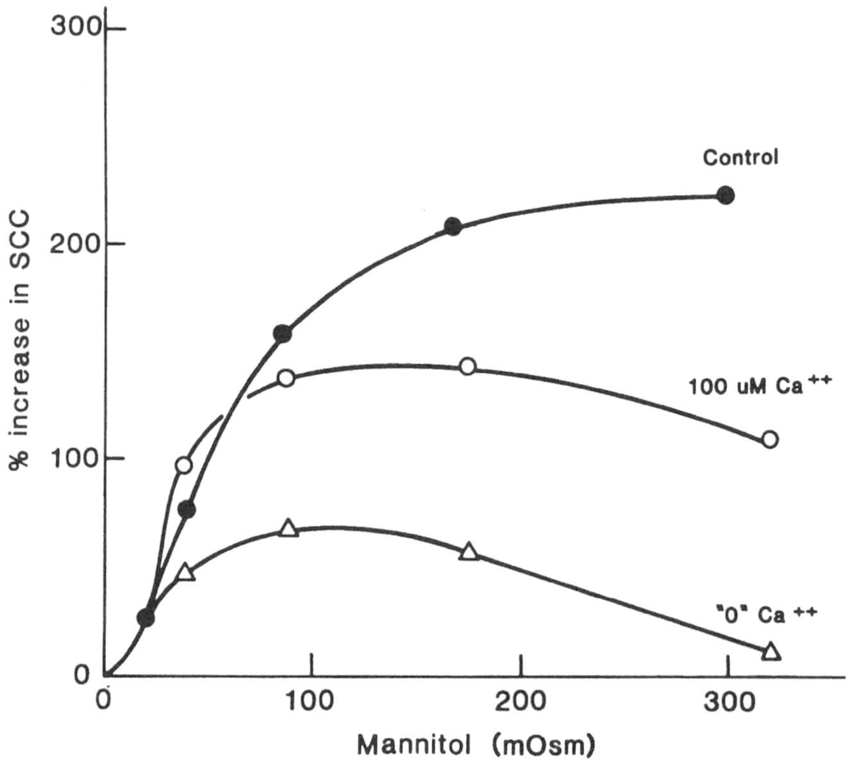

Fig. 6.5. Dose-response curves of chloride current versus increased osmolarity on the basolateral side. Symbols are: (●) control; (○) 100μM Ca^{++}; (Δ) "0" Ca^{++}. Each point is the mean of 6 experiments. The system is extremely sensitive to hypertonicity; 12.5 mOsm induced more that 20% increase in current output. Maximal increases in chloride secretion are observed above 200 mOsm on the basolateral side. Reduction in Ca^{2+} concentration in solutions bathing both sides of the operculii produces drastic reduction on the hypertonic response. Data are reprinted with permission from the *Journal of Membrane Biology*.

6.2.3 Dose response curves for osmolarity versus chloride current and an effect of Ca concentration

In Fig. 6.5 we can observe the effects on current of increased levels of osmolarity ranging from 50 to 200 mOsm in the basolateral side of opercular preparations. The curve describes a saturable event and reveals that small changes in osmolarity are able to produce dramatic changes in chloride secretion. This translates into small changes of salt concentration in the fish plasma produce large changes in chloride secretion in the gill. In fact 12.5 mOsm increase in osmolarity, a change of 4.1 % above the normal osmolarity of 300 mOsm produces an increase in secretion of more than 20% in the chloride current. If we return to Fig. 6.2 where the increase in plasma osmolarity is shown, the peak of the osmolarity increase is reached after 10 hrs at 65 mOsm above an initial plasma osmolarity of 280 mOsm. This means that a 23% increase in plasma osmolarity can produce an increase of almost one and half times in chloride transport. It takes the fish 73 hours to return the plasma to the original osmolarity, therefore 65 mOsm, which are equivalent to 32.5 mM of chloride, are pumped out in 73 hours, a rate or flux of 445 μEq/h. In the short circuited opercular epithelium the net chloride transport is 4.79 μEq/h per square centimeter of surface. This value of 445 μEq per h is approximately 9 to 10 times greater than the net flux in the opercular epithelium. Therefore one of the possibilities is that the area for secretion in the gills is 9 to 10 times greater than the area of the opercular epithelium. We know that the opercular epithelium has an approximate area of 1 sq cm. with 10,000 chloride cells (Zadunaisky, 1984). This means that 10 sq. cm could be the secretory area occupied by the chloride cell epithelium in the gill containing from 90,000 to 100,000 chloride cells.

Ca ions are important for the process of secretion of chloride as well as for the effect of osmolarity. In Fig. 6.5 the changes in the control curve of osmolarity versus chloride secretion produced by the reduction of Ca^{2+} in the medium from 1 mM to μM and to nominal "zero" calcium are shown. There is a less rapid increase in chloride secretion with osmolarity and a reduced maximal steady state level after 200 mOsm. Ca^{2+} is needed for the coupling between tonicity and secretion, as well as for the normal function of the cytoskeleton in all processes of cell expansion or shrinkage. The actual mechanisms of the effect of reduced Ca^{2+} are not clear, which is also true in several other transporting epithelia.

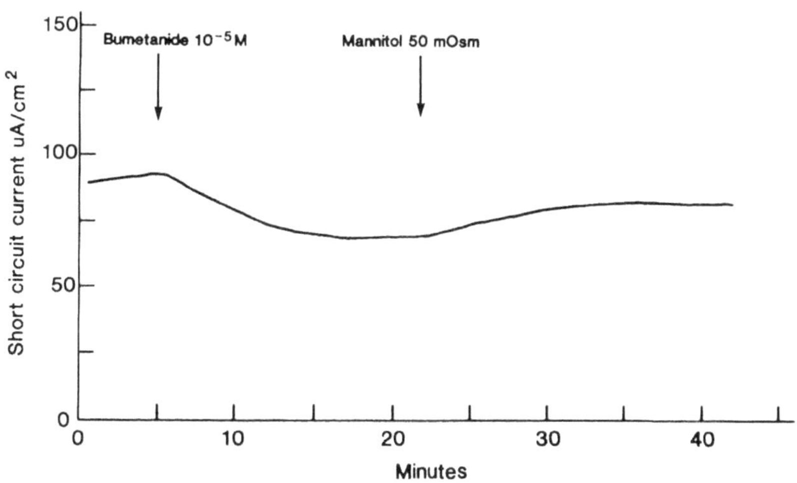

Fig. 6.6 Lack of response to basolateral hypertonicity of 50 mOsm induced with mannitol, after addition of 10^{-4}M bumetanide to the basolateral side. The $Na^+K^+2Cl^-$ cotransporter is most probably involved in the activation by cell shrinkage. Reproduced with permission from J. Membrane Biol.

6.2.4 Blocking the 2Cl,Na,K cotransporter eliminates the response to mannitol

The inhibition of the basic chloride current with bumetanide, a rather specific blocker of the 2Cl,Na,K cotransporter consists in a slow reduction in current that in about 2 hours is radically reduced. In Fig. 6.6 it is shown that addition of 50 mOsm of mannitol to the basolateral side of an isolated opercular preparation inhibited with bumetanide does not produce the expected increase in chloride current. This means that without the basic entry of chloride ensured by the 2Cl,Na,K cotransporter the whole effect of shrinkage and activation of chloride channels is lost. Therefore this cotransporter is activated during the addition of mannitol or is part of the basic sequence of events leading to more chloride secretion.

6.2.5 Presence of the Na^+/H^+ exchanger and the consequences of its inhibition on the response to mannitol

Does the Na^+/H^+ exchanger participate in the response to hyperosmolarity by the chloride cells? Or in other words is there a pH change that activates this exchanger during the increase of osmolarity and subsequent increase in chloride secretion? First we tested for the presence of the Na^+/H^+ exchanger; 15 mM of ammonium chloride were added to the basolateral side and it produced a decrease in current associated with the acidification of the chloride cells (see Boron and De Weer 1976). Then the basolateral side was washed out with regular medium eliminating the NH_4Cl. At this point the current increased again.

It has been demonstrated in other systems that the activation of the Na^+/H^+ exchanger by the displacement of the intracellular pH, is compensated by the Na^+/H^+ exchanger and the rate of rise of the current to normal is a function of the rate of exchange of H ions out of the cell for Na ions into the cell. These experiments in opercular membranes are shown in Fig. 6.7 A and B It can be appreciated that ammonium chloride does not affect the current of the operculum from the apical side, the effect is only obtained in the basolateral side. Thus a Na^+/H^+ exchanger is present in the basolateral side of the chloride cells. It can be inhibited or activated by a number of agents. The specific inhibitors are amiloride and its analogues. The action of amiloride at high concentration is shown in the last column of Fig. 6.7 B as well as the stimulating effects of epithelial growth factor (EGF), phorbol ester (PMA) and IBMX. The exchanger is present and responds as expected.

Does it participate in the increase in current produced by the hypertonicity? Fig. 6.8 shows that it does because addition of amiloride at a high dose before the test for hypertonicity with 50 mOsm of mannitol completely suppresses the expected increase in chloride current. Therefore the integrity of the Na^+/H^+exchanger is essential for the hypertonic response.

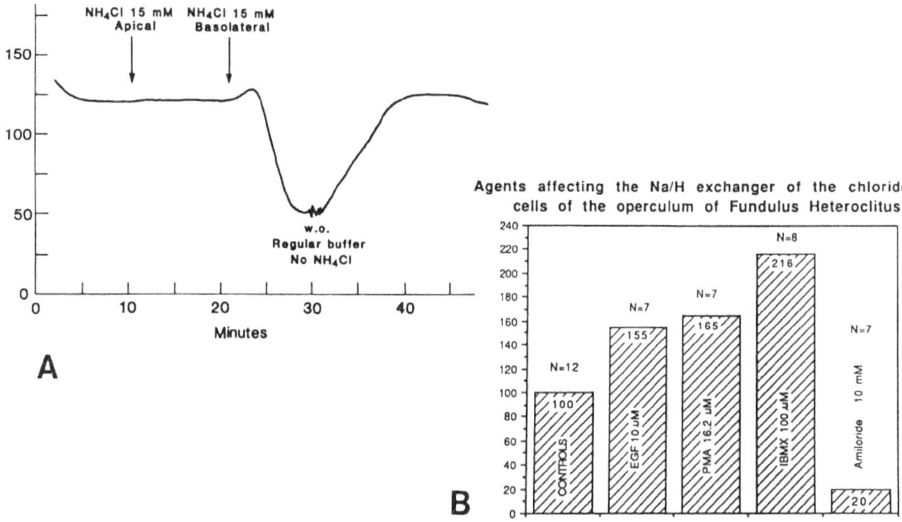

Fig. 6.7 (A) + (B) (A) Na^+/H^+ exchanger. Action of NH_4Cl on the chloride current of isolated chloride-cell-rich opercular epithelium. Acidification of intracellular pH and return to normal produce parallel effects on the chloride current. The influence was only found on the basolateral side. (B) Stimulatory effects on the recovery of the chloride current after washout of NH_4Cl from the basolateral side. IBMX, phorbol ester, and EGF have stimulatory effect while amiloride inhibits the recovery. A Na^+/H^+ exchanger is present and activated or blocked from the basolateral side.

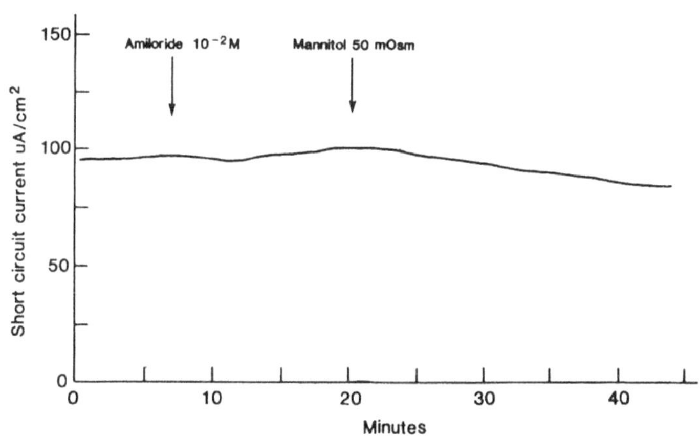

Fig. 6.8 Amiloride block of the chloride current increase expected with hypertonicity of the basolateral side. The Na^+/H^+ exchanger is activated or participates in the hypertonic response of the chloride secretion.

We tested amiloride at several concentrations and analogues of this drug that are more potent and can be used at somewhat lower concentrations. The results are shown in Fig. 6.9A and B. Amiloride at 10^{-2} M completely blocks the effect of 50 mOsm of mannitol, this is shown in the last bar of Fig. 6.9 A. At 10^{-4} it produces 20% inhibition and at 10^{-3}, 80% inhibition. The relative effects of amiloride and two of its analogues, HMA and EIPA both tested at 10^{-4} M show that the analogues induce 50% inhibition of the chloride current activation by mannitol against the effect of amiloride of about 20%.

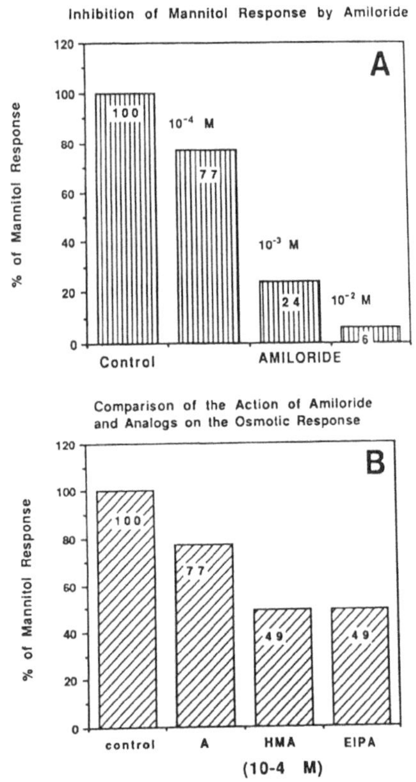

Fig. 6.9 **(A) + (B)** (A) Relative inhibition of the osmotic response found with different concentrations of amiloride in isolated opercular epithelia. Blocking the Na^+/H^+ exchanger elimnates the ixcncrease of the chloride current induced by basolateral hypertonicity. Bars are mean of seven experiments each. The numbers inside the bars indicate the percentage of the response remaining after treatment. (B) Relative inhibition of the current increase produced with mannitol by amiloride and its analogs HMA and EIPA at 10^{-4}M concentration. Bars are mean of seven experiments each. Numbers inside them indicate percentage of current stimulation remaining after drugs.

6.2.6 Presence of the Na^+/HCO_3^- exchanger and its lack of participation in the hypertonic response.

The presence of the chloride bicarbonate exchanger in chloride cells was suspected because of the dramatic effects of the bicarbonate ion on the chloride current (Zadunaisky 1984). In fact in Fig. 6.10 A and B the complete elimination of CO_2 and bicarbonate, replaced by a HEPES buffer, produces a reduction of current to about 1/3 of its original value, that recovers when the CO_2 and HCO_3^- are replaced In Fig. 6.10 B it is observed that the highly specific inhibitor of this exchanger, DIDS, reduces the current also to 1/3 of its original value in the presence of CO_2 and HCO_3 and their removal has no further effects. There is therefore a system of transport of chloride into the cells and extrusion of HCO_3^- on the basolateral side of the chloride cell.

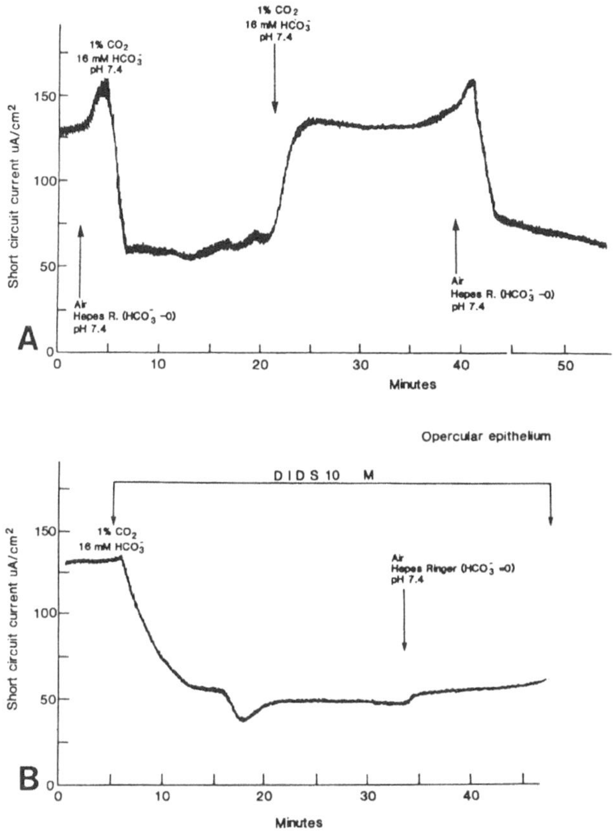

Fig. 6.10 (A) + (B) Cl^-/HCO_3^- exchanger. (A) The removal of HCO_3^- and CO_2 keeping the pH of the bathing solutions at 7.4 with HEPES buffer, produced remarkable reduction of chloride current. Recovery occurs when HCO_3^-, CO_2 rich medium is readmitted. The first part of the sequence is repeated. (B) Inhibition with DIDS of the chloride current in the presence of HCO_3^- and CO_2. Note that HEPES Ringer, free of HCO_3^- or CO_2 do not elicit further reduction of the current. The effects of DIDS or the removal of bicarbonate had proportionally similar inhibitory action.

Interestingly this exchanger in not involved in the hypertonic activation of chloride current by 50 mOsm of mannitol. It can be seen in Fig. 6.11 that after reduction of the current by DIDS, addition of mannitol produces the full effect expected. We conclude that the Cl^-/HCO_3^- exchanger is not participating in the hypertonic response.

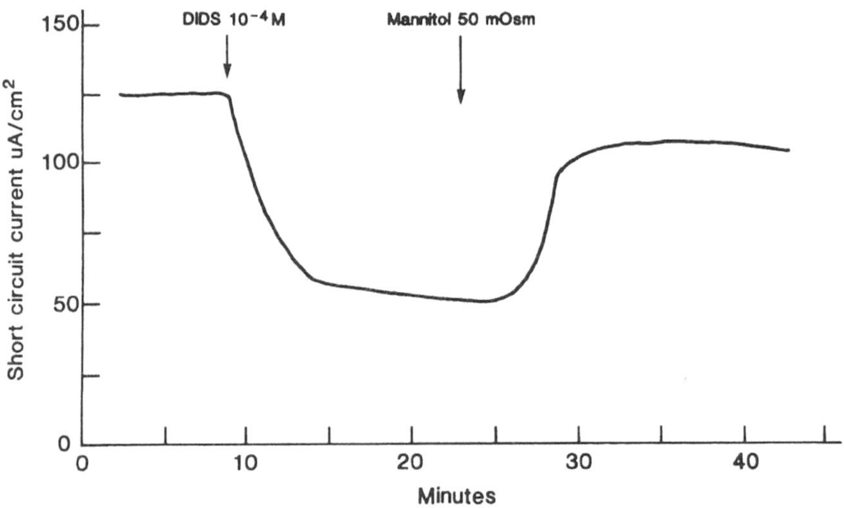

Fig. 6.11 Blocking of the Cl^-/HCO_3^- exchanger with DIDS did not affect the increase in chloride current expected when the tonicity of the basolateral side of isolated opercular epithelium is increased with mannitol. The Cl^-/HCO_3^- exchanger does not participate in the hypertonic response.

6.2.7 Blocking Cl channels with DPC

The integrity of the apical chloride channels is essential for the maintenance of the basic chloride current of the chloride cells as well as for the response to the changes in tonicity in the medium. In fact Fig. 6.12 shows the effect of the chloride channel blocker DPC added to the basolateral side of an opercular preparation. The current drops dramatically. After 20 minutes when the current had dropped below 50% we added 50 mOsm of mannitol. No change in the rate of drop of current is detected, indicating that the chloride channels are directly involved in the hypertonic response.

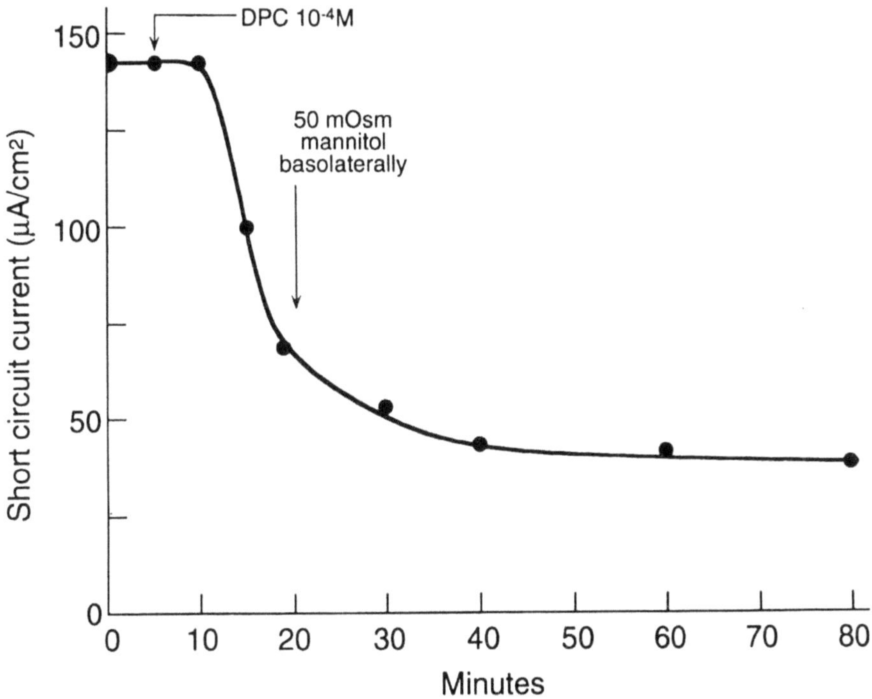

Fig. 6.12 Inhibitory action of the chloride channel blocker DPC at 10^{-4}M on the chloride current of an operculum and lack of response to basolateral hypertonicity with 50 mOsm mannitol. The chloride channel efficacy is needed for the osmolarity to induce its shrinking and stimulatory effect of chloride secretion.

6.2.8 Quantitative microscopy of chloride cells, an imaging analysis

In order to evaluate the response to hyper and hypotonicity of chloride cells, opercular epithelia were mounted in an *ad hoc* flat chamber that permitted perfusion of each side of the preparation during observation in an inverted microscope fitted with refraction optics (Nomarsky). The chloride cells were observed, and optical sections were performed. The images were stored in a computer hard disc, and areas and cell volume measured by a computerized method based on planimetry (Zadunaisky et al. 1995). The epithelia were perfused with solutions with a 20% increase or decrease in (60 mOsm) tonicity by addition of mannitol or dilution. Operculii obtained from fish adapted to fresh water or to sea water were used. Table 6.1 shows the results of four series of experiments.

Tabe 6.1 Effects of 20mOsm of mannitol on the cell volume of chloride cells

Adaptation	Side	Cell volume μm^3	Minimal volume μm^3	%	Shrikage % of control
Fresh Water	Apical $N = 6$	1,265 ± 189	1,299 ± 118	81.4* ± 3.8	19.9 ± 5.1
	Basol. $N = 5$	1,750 ± 208	1,414 ± 165	80.8* ± 4.0	17.8 ± 4.9
Sea Water	Apical $N = 6$	1,537 ± 222	1,570 ± 248	103.0 ± 6.0	1.2 ± 3.0
	Basol. $N = 5$	1,660± 191	1,328 ± 131	80.0* ± 3.6	21.0 ± 6.1

Table 6.1 Mannitol was added to the apical or the basolateral side perfusion of sea water or fresh water adapted operculiiij of *F. heteroclitus*. Mean cell volumes are followed by ±SE. *N* indicates number of cells in which the volume was measured by planimetric methods on a minimum of six and maximum of nine optical sections obtained at intervals of 2.5 µm. (*)Statistically significant differences. The exposure of the apical membrane of sea-water -adapted cells did not produce significant cell volume change indicating low water permeability. Apical membrane of fresh water and basolateral of both sea-water and fresh-water-adapted produced expected volume shrinkages of approximately 20%. Data available for the return from hypertonic to isotonic solutions produced results that permit a similar interpretation.

In the case of the sea water adapted fish, increased tonicity on the basolateral side produced approximately 20% reduction in cell volume, as expected from a permeable cell membrane without restrictions to water movements. The apical side of the sea water adapted fish appears to be much less permeable to water as no change in volume was observed with hypertonicity. In fresh water adapted fish the basolateral side responded as expected with reduced volume also of close to 20% with a hypertonicity of 20% above the medium. The only suprise was the action of the mannitol on the apical side of fresh water adapted fish. There was a 19.9% change in volume of chloride cells, indicating that the apical side after more than a month of adaptation to full fresh water became more permeable to water, or as permeable as the basolateral side. It is possible that in real life *Fundulus heteroclitus* moving from fresh water to sea water can detect the high tonicity through the apical membranes. However, this fish lives normally in brackish water and moves to sea water and back to the estuary daily, a time period that probably does not permit the adaptation we observed in specimens kept for prolonged periods in full fresh water.

6.2.9 Effects of hypotonic solutions on the chloride current

Up to this point we have referred mostly to the action of hypertonicity on chloride secretion on opercular membranes rich in chloride cells. What is the effect of reducing the tonicity that is imitating the transition from sea water to fresh water and testing the effects of swelling, instead of shrinkage of chloride cells. A summary of preliminary data (Zadunaisky, Au, Bala and Colon, in preparation) indicate the following : 1) reduction of tonicity has effects on the chloride current only on the basolateral side, 2) the system is extremely sensitive to small changes of osmolarity, 5-10 mOsm produce drastic reduction of the chloride current. 3) the effects of dilution are mostly due to osmolarity reduction and only partly to specific reduction in chloride ion concentration. 4) utilizing highly specific inhibitors to test the transport, cotransport and exchange of ions we find that the Cl^-/HCO_3^- exchanger is essential for the response to hypotonicity. 5) the Na^+/H^+ exchanger is not needed 6) the 2Cl,Na,K cotransporter is essential for the effects of reduced osmolarity 7) the integrity of the chloride channels is important 8) part of the reduction in chloride secretion produced by hypotonicity, is due to the opening of a K^+ channel in the basolateral side of the opercular epithelium.

6.3 Conclusions

An overall interpretation of the information and its analysis provided in this paper indicate that imitation of the transient changes in plasma osmolarity of fish that can make the transition to extreme levels of salinity, utilizing the opercular epithelium shows chloride cells respond with shrinkage to increases in osmolarity and swelling to reductions of osmolarity. The shrinkage of the chloride cell activates the 2Cl,Na,K cotransporter, the Na^+/H^+ exchanger and the chloride channels of the apical side of the cells. Calcium is required for the full expression of the increase in chloride secretion in the transition to high salinity. In the reverse direction towards lower salinities, the slight reduction of plasma osmolarity, extrapolating from the response of opercular membranes, is enough to reduce the chloride secretion by swelling the cells, activating the Na^+/HCO_3^- exchanger, the 2Cl,Na,K cotransporter and probably closing chloride channels in the apical membrane. The overall effect of hypotonicity is facilitated by the opening of a K^+channel in the basolateral side. This hypothesis of the nature of the rapid signal for adaptation, is based in the experiments presented here. The secondary adaptation, involving the proliferation or apoptosis of chloride cells, is obviously an area of future research.

Acknowledgments:

Dr. Zadunaisky was a Scholar of the Fogarty International Center of the National Public Health, National Institutes of Health, Bethesda MD, USA during the experimental work discussed in this review article. The collaboration of Kenneth S. Spring, David Croft, Elizabeth Colon and Michael Bala at NIH, NYU and MDIBL is gratefully acknowledged. The work was supported by NIH grant EY01340 to J.A.Z. I am grateful to Drs. F.B. Eddy and N. Hazon for their kind invitation to participate in this well deserved Symposium in Honor of William Potts, scholar extraordinaire and scientific father to a generation of marine biologists, at the University of Lancaster, England.

References

Bevelander, G. (1936). Branchial glands in fishes. *J. Morphol.*59, 215-224

Boron, W.F. and De Weer, P. (1976). Intracellular pH transients in squid giant axon caused by CO_2,, NH_3 and metabolic inhibitors. *J. Gen. Physiol.* 67, 91-112

Burns, J. and Copeland, D.E. (1950). Chloride excretion in the head region of *Fundulus heteroclitus. Biol. Bull. (Woods Hole).* 99, 381-385

Degnan, K.J., Karnaky, K.J. and Zadunaisky, J.A. (1977). Active chloride transport in the *in vitro* opercular skin of a teleost (*Fundulus heteroclitus*) a gill-like epithelium rich in chloride cells. *J. Physiol.* 294, 155-191

Degnan, K.J., and Zadunaisky, J.A. (1980a). Passive sodium movements across the opercular epithelium: The paracellular shunt pathway and ionic conductances. *J. Membr. Biol.* 55, 175-185

Degnan, K.J., and Zadunaisky, J.A. (1980b). Ionic contributions to the potential and current across the opercular epithelium. *Am. J. Physiol.*. 238, R231-R239

Foskett, J.K., Hogsdon, C.D., Turner, T., Machen, T.E. and Bern, H.A. (1981). Differentiation of the chloride extrusion mechanisms during seawater adaptation of the teleost fish, the cichlid *Sarotherodon mossambicus. J. Exp. Biol.* 93, 209-224

Harvey, B.J., Thomas, S.R. and Ehrehnfeld, J. (1988). Intracellular pH controls cell membrane Na^+ and K^+ conductances and transport in frog skin epithelium. *J. Gen. Physiol.* 92, 767-791

Hoffmann, E.K., and Simonsen, L.O. (1989). Membrane mechanisms in volume and pH regulation in vertebrate cells. *Physiol. Rev.* 69, 315-382

Karnaky, K.J., Kinter, L.B., Kinter, W.B., and Stirling, C.E. (1976). Teleost chloride cell. II. Autoradiographic localization of gill NaKATPase in killifish *Fundulus heteroclitus* adapted to low and high salinity environments. *J. Cell. Biol.* 70, 157-177

Keys, A.B. (1931). The heart-gill preparation of the eel and its perfusion for the study of a natural membrane *in situ. Z. Vergl. Physiol.* 15, 352-363

Keys, A.B. and Willmer, E.N. (1932). "Chloride secreting cells" in the gills of fishes with special referenceto the common eel. J. Physiol. (Lond.) 76, 368-378

Maetz, J. and Bornacin, M. (1975). Biochemical and biophysical aspects of salt excretion by chloride cless in teleosts. *Fortschr Zool.* 23, 322-362

Marshall, W.S. (1977). Transepithelial potential and short-circuit current across the isolated skin of *Gillichthys mirabilis* (Teleostei: Gobiidae), acclimated to 5% and 100% seawater. *J. Comp. Physiol. B. Biochem. Syst. Environ. Physiol.* 114, 157-165

Marshall, W.S., Bryson, S.E., Midelfart, A. and Hamilton, W.F. (1995). Low-conductance anion channel activated by cAMP in teleost Cl^- -secreting cells. *Am. J. Physiol.* 268, R963-R969

Scheide, J.I. and Zadunaisky, J.A. (1988) Effect of atriopeptin on isolatedopercular epithelium of *Fundulus heteroclitus.* Am. J. Physiol. 254, R27-R32

Smith, H.W. (1930). The absorption and excretion of water and salts by marine teleost. *Am. J. Physiol.* 93, 485-505

Rowing, G.M., and Zadunaisky, J.A. (1978). Inhibition of chloride transport by acetylcholine in the isolated opercular epithelia of *Fundulus heteroclitus*. Presence of a muscarinic receptor. *Bull. Mt. Desert Isl. Biol. Lab.* 18, 101-104

Zadunaisky, J.A. (1984). The chloride cell: The active transport of chloride and the paracellular pathways. In *Fish Physiology* (vol XB) (eds. W.S. Hoar and D.J. Randall), pp 129-176. New York: Academic Press

Zadunaisky, J.A., and Degnan, K.J. (1976). Passage of sugars and urea across the isolated retina pigment epithelium of the frog. *Exp. Eye Res.* 23, 191-196

Zadunaisky, J.A., Curci, S., Schettino, T. and Scheidi, J.I. (1988). Intracellular voltage recordings in the opercular epithelium of *Fundulus heteroclitus*. *J. Exp. Zool.* 247, 126-130

Zadunaisky, J.A., Cardona, S., Au, L., Roberts, D.M., Fisher, E., Lowenstein, B., Cragoe, E.J. Jr, Spring, K.R. (1995). Chloride transport activation by plasma osmolarity during rapid adaptaion to high salinity of *Fundulus heteroclitus*. *J. Membr. Biol.* 143, 207-217

The Use of Modern Microscopical Techniques for the Study of Fish Gill

A.J.H. van der Heijden[1] and I.J. Morgan[2]

[1] Department of Animal Physiology, University of Nijmegen, Toernooiveld, 6525 ED Nijmegen, The Netherlands

[2] Department of Biology, McMaster University, 1280 Main Street West, Hamilton, Ontario, Canada L8S 4K1

7.1 Introduction

The branchial epithelium of the fish gill consists of at least four different cell types: pavement cells, chloride cells, mucus cells and neuroepithelial cells (Laurent, 1984). Of these, the pavement cells (or respiratory cells when located on the lamellae) and the chloride cells, also called ionocytes or mitochondria-rich (MR-) cells, are thought to be directly involved in gas exchange, ionoregulatory and acid/base-regulatory processes. Although gill cells have been extensively studied, in particular chloride cells (reviewed by Zadunaisky, 1984; Pisam and Rambourg, 1991; Marshall, 1995; McCormick *et al.*, 1995), important questions about these cells are as yet unanswered. For example, on the basis of ultrastructural and biochemical studies, the chloride cells are generally thought to be responsible for ion transport across the branchial epithelium of fish gills. However, recent evidence suggests the presence of proton pumps and Na^+/H^+-exchanger in the respiratory cells, indicating a possible role in Na^+-uptake in freshwater fish (Lin and Randall, 1995; Sullivan *et al.*, 1995). To define the ion transporting function of chloride and respiratory cells direct *in situ* measurements of ion transport, membrane potentials or intracellular ion concentrations are needed. Up to now intracellular electrodes could not be used for these measurements because of the complex structure and vascularization of the gills, which hinder good access to individual cells.

Due to these limitations, conclusions about the function of chloride cells are mostly drawn from correlations between ion fluxes or biochemical data and estimated numbers of chloride cells. To make a reliable correlation between these measurements, quantitative data on *functional* chloride cells are required. But how does one define a functional chloride cell, and how are these cells quantified?

In this paper we will discuss microscope techniques to study gill cells *in situ*. We will show how the confocal laser scanning microscope (CLSM) can be used in combination with fluorescent probes to identify and quantify chloride cells in freshly isolated, unfixed gills. With X-ray microanalysis (XRMA) the total concentration of elements can be measured in individual cells of frozen tissue. This technique was used to measure intracellular elemental concentrations in pavement and chloride cells in response to changes in ion concentrations in the ambient water and chemicals that are known to affect ionoregulation.

7.2 Quantification of Chloride Cells

The chloride cell is characterized by a number of ultrastructural features (Laurent, 1984), which are unique to ion transporting cells (Berridge and Oschman, 1972). Microscopic methods for the quantification of chloride cells use the distinctive features of these cells. However, within a population of ion transporting cells functional differences may exist, as has been shown for the mitochondria-rich cells in the collecting duct of the mammalian kidney (Brown *et al.*, 1985; Schuster *et al.*, 1986). In fish gills, electron microscopical observations have revealed that the chloride cell population in this epithelium is not homogeneous (see below), and this has to be taken into account when quantitative data are used for correlative studies.

7.2.1 Chloride Cell Subtypes

To date only morphological and topological criteria have been advanced for subtyping the chloride cell population. In early studies two types have been distinguished; chloride cells and accessory cells, the latter considered as either developing or degenerating chloride cells (Coleman *et al.*, 1977; Sardet *et al.*, 1979; Hootman and Philpott, 1980).

Pisam and co-workers (1987; 1993) distinguished two different subtypes (α and ß) of chloride cells in guppy and tilapia based on location, shape and cytoplasmic staining. They observed that this structural diversity is related to the ionic composition of the water, with α-cells dominating in seawater and ß-cells being the dominant freshwater type.

Scanning electron microscopical (SEM) examination of gills of salmonids have revealed differences in surface morphology (smooth or with microvilli or microridges) displayed by chloride cells in the same filament (Perry and Laurent, 1993). Whether the structural differences reflect functionally different chloride cells is not known.

Wendelaar Bonga and van der Meij (1989) concluded from their study in tilapia that the chloride cell develops through accessory and immature stages to the mature stage and eventually disappears through physiologically controlled cell death (apoptosis) or death induced directly by extrinsic factors (necrosis). Within these five developmental stages of the life cycle of the chloride cell, only the "mature" stage shows all features anticipated to define it as a functional ion transporting cell. Firstly, mature cells have many mitochondria which provide the energy required for active transepithelial ion transport. Secondly, the cells have an extensive tubular network that is continuous with the basolateral membrane and where ion transporting enzymes such as Na^+/K^+-ATPase, Ca^{2+}-ATPase and Na^+/Ca^{2+}-exchangers are located (Karnaky *et al.*, 1976; Flik *et al.*, 1985; Verbost *et al.*, 1994). Thirdly, they contact both the water via an apical membrane and the extracellular fluid with the basolateral membrane. Accessory chloride cells in tilapia are considered functionally inactive "replacement" cells, because they lack a fully developed tubular system and are not necessarily in contact with the water. Immature cells are structurally intermediate between accessory cells and mature cells. Apoptotic and necrotic chloride cells rapidly lose contact with the water and thus are unlikely to be functionally active.

In the following sections we will use the terminology for different chloride cell "subtypes" as proposed by Wendelaar Bonga and co-workers; with the so-called mature chloride cells representing functional ion transporting cells. As mentioned above, this is a morphological criterium. Direct measurements of ion transport in chloride cells are scarce (Foskett and Scheffey, 1982; Marshall *et al.*, 1995), and are mostly not combined with subtyping of the cells.

7.2.2 Techniques for Characterization and Quantification of Chloride Cells

A number of microscopical techniques has been applied for the quantification of chloride cells. These include the use of transmission electron microscopy (TEM), scanning electron microscopy (SEM) and light microscopy (LM). Each of these techniques has its own particular advantage or drawback.

The TEM has a high resolution and identification of the subtypes was based on this technique. However, the preparation and quantitative analysis of chloride cells and their subtypes in an architectural complex organ like the gills is very laborious.

With SEM one can examine the external aspects of the branchial epithelium. TEM analyses show that mature chloride cells appear as two morphologically distinct types, viz. cells with a convex apex and cells with a concave apex (called crypt or pit). SEM has been used to quantify the size and density of the chloride cell apex, and it was found that changes in the product of these two parameters (the so-called "chloride cell fractional area") was positively correlated to changes in ionfluxes (Perry *et al.*, 1992*a* and 1992*b*). However, in some species the similarity between apical crypts and mucus cell pores may impair a reliable identification and quantitation of chloride cells with SEM (van der Heijden *et al.*, in press). Further it has to be kept in mind that fixation or drying procedures for SEM can cause shrinkage of the tissue leading to an overestimation of absolute numbers (van der Heijden *et al.*, in press). Another limitation of this method is that it gives no information about the total population of chloride cells and the relative densities of the different subtypes for which only TEM analysis would suffice.

At the LM level, chloride cells can be distinguished by their affinity for a staining procedure involving a mixture of zinc iodide and osmium tetroxide ("ZIO")(Garcia-Romeu and Masoni, 1970; Avella *et al.*, 1987). However, macrophages and other leucocytes have also been found to be ZIO-positive (Dagdeviren *et al.*, 1994). In the branchial epithelium, macrophages are present to eliminate apoptotic chloride cells, and their number increases with the increased occurrence of apoptotic cells, for example when fish are transferred to seawater, to acidified water or water containing toxic substances (Wendelaar Bonga and van der Meij, 1989; Wendelaar Bonga *et al.*, 1990; Pratap and Wendelaar Bonga, 1993).

Also enzyme-cytochemical methods can be applied for LM. Radioactive or fluorescent derivatives of specific inhibitors such as ouabain (Karnaky *et al.*, 1976; McCormick, 1990), and antibodies with specific affinity for Na^+/K^+-ATPase (Witters *et al.*, 1996) have been used to visualize chloride cells through their high Na^+/K^+-ATPase content. Assuming that a probe binds to the active form of the enzyme only, as is the case with anthroylouabain (Moczydlowski and Fortes, 1980), a distinction can be made between cells with active and inactive enzyme and this may reflect the activity of the cell.

However, additional fluoresent staining procedures are now available for conventional fluorescence microscopy as well as confocal laser scanning microscopy (CLSM). Vital stains for mitochondria such as DASPEI and DASPMI (Bereiter-Hahn, 1976) are used for the identification of chloride cells, which contain an abundance of mitochondria. Since not only mature chloride cells, but also accessory, immature and perhaps also degenerating chloride cells may have many metabolically active mitochondria, most or all subtypes are labelled by this probe.

Our studies have shown that certain plant lectins can be used to identify mature chloride cells. Lectins bind to glycoproteins that constitute the glycocalyx of the apical plasma membrane. The type of glycoproteins that each cell type possesses on its surface may be related to its function. Indeed, ultrastructural studies of the glycocalyx of gill cells showed that the thickness (Powell *et al.*, 1994) and composition (Pisam *et al.*, 1980) of the cell coat of chloride cells is different from that of pavement cells, mucocytes or adjacent (accessory) cells. It appeared possible to specifically label the apical membrane or glycoproteins secreted in the crypt of MR-cells using the jack bean lectin, Concanavalin-A (Con-A) as a tool (see also next paragraph). In the mammalian kidney, peanut agglutinin has been used to specifically label the apical membrane of MR-cells (Brown *et al.*, 1985; Schuster *et al.*, 1986).

Although the use of fluorescent probes has allowed the direct and instantaneous examination of flat MR-cell containing epithelia such as the opercular membrane and gill cell suspensions, conventional fluorescence microscopy has the disadvantage that it is not suitable for the study of the multilayered and complex epithelium of the gill. Compared to conventional fluorescence microscopy, CLSM offers superior imaging properties: resolution is improved, out-of-focus blur is reduced and microscopy in three dimensions can be performed. In the next paragraphs we will give two examples of how visualization of MR-cells in the branchial epithelium of fish gills by CLSM has extended our knowledge of the gills.

7.2.3 Development of Functional Branchial Chloride Cells in Larvae and Juveniles

The first example concerns the timing of the development of functional chloride cells in the fish gill. Embryos and larvae of teleosts are able to maintain hydromineral homeostasis, even when their gills and kidneys are still poorly developed (Alderdice, 1988). Extrabranchial chloride cells with apical crypts have been found in the body skin and in the yolk-sac epithelium in freshwater tilapia embryos (already 48 hrs after fertilization, Hwang *et al.*, 1994) and in larvae (1 day after hatching, Ayson *et al.*, 1994). These extrabranchial chloride cells begin to disappear during the post-yolk sac stage (10 days after fertilization) and this coincides with the start of gill development (Hwang *et al.*, 1994). This correlation has led to the assumption that these extrabranchial chloride cells are the main site of ionoregulation in early developmental stages before the gills become functional. However, little is known about the development of branchial chloride cells or when they become operational.

We have applied a combination of morphological, histochemical and biochemical methods to identify and quantify functional MR-cells in the gills of larval and juvenile freshwater tilapia, *Oreochromis mossambicus* (Li *et al.*, 1995). We used three criteria to identify functional MR-cells; the presence of an abundance of mitochondria, an apical crypt and a high Na^+/K^+-ATPase content. To identify MR-rich cells with the CLSM the fluorescent probe DASPMI was used with Con-A fluorescein conjugate to label the apical crypt. The *in vivo* co-labelling procedure for the gills of these young developing fish (post-hatching age of 3, 10 and 24 days) for CLSM has been described by Li *et al.* (1995).

Upon examination of double-labelled gill arches, the filamental epithelium showed up as a single layer of DASPMI-positive cells (Fig. 7.1). Fluorescein conjugated Con-A labelling was found associated with practically all of these cells. When consecutive optical sections from a plane parallel to the mucosal surface of the double-labelled gill filaments and to a depth of one cell layer are projected orthogonally (Fig. 7.2.), it can be seen that the two labels are confined to single cells.

TEM analysis confirmed the presence of numerous mitochondria and apical crypts in gill chloride cells of these young developing fish. These cells also exhibited an extensive tubular system, where ion-transporting enzymes are located. In the CLSM, exposure of the gills to 5 μM of the fluorescent probe anthroylouabain (which binds to the extracellular binding site of Na^+/K^+-ATPase) resulted in the appearance of many fluorescent cells, which resembled the DASPMI-positive cells on the basis of size, morphology and location (see Li *et al.*, 1995). Simultaneous exposure of the gills to a much higher concentration of ouabain (1 mM) completely abolished the fluorescent signals, demonstrating that chloride cells at this early stage of development contain functional ouabain-binding sites. Biochemical determination of Na^+/K^+-ATPase activity in gill homogenates of 10-day-old larvae showed a high specific activity (1558 ± 366 μmol $P_i.h^{-1}.mg^{1}$). This value was more than 380 times higher than the specific activity found in adults (4.04 ± 0.63 μmol $P_i.h^{-1}.mg^{-1}$). From these data we conclude that chloride cells in the gills of larvae and juvenile freshwater tilapia may be qualified as functional chloride cells participating in ion transport, even before the branchial lamellae start to develop.

Although chloride cells are mainly located in the gills, they may also occur in extrabranchial areas, in particular the inner opercular epithelia and the skin, of many fish species. In larvae, most extrabranchial chloride cells are located in the skin covering the yolk sac. To establish which of the two groups of chloride cells - the branchial or the extrabranchial - are the most important site of ion transport in the larvae, MR-cell numbers in gills and skin were compared. The density of MR-cells in the gills ($n.mm^{-2}$ ± S.D.) was 4630 ± 489 in 3-day old fish, 6280 ± 271 in 10-day old fish and 6600 ± 219 in 24-day old fish. The total MR-cell population in the gills of 3-day old fish was estimated to consist of at least 20400 cells (see Li *et al.*, 1995), whereas we calculated a maximum number of 1500 MR-cells in the yolk sac epithelium (based on a maximum MR-cell density of 500 mm^2, and a total yolk sac surface of 3 mm^2 as measured by Ayson *et al.*,

1994). Clearly, the MR-cell population in the gills outnumbers that in the skin indicating that the branchial cells play the major role in active ion transport already during the early developmental stages.

7.2.4 Chloride Cell Density in Adult Fish Adapted to Fresh water or Seawater

The second example concerns the use of CLSM for the determination of the distribution, cell size and number of chloride cells in adult fish, for which tilapia adapted to either fresh water or seawater were studied. Whereas the chloride cells are mainly involved in the active ion absorption for growth (Ca^{2+}) and homeostasis (Na^+, Cl^-) in freshwater fish, they are mainly involved in secretion of the surplus of Na^+ and Cl^- that are absorbed by the intestine in seawater fish (Foskett *et al.*, 1981; Maetz, 1974). Ca^{2+} is actively absorbed by the gills in fresh water as well as in seawater (see review by Flik *et al.*, 1996*b*). Euryhaline fishes, like the Mozambique tilapia, can adapt to different salinities and thus are able to reverse Na^+ and Cl^- fluxes via their gills. Assuming that active branchial transport of these ions exclusively takes place via functional chloride cells (which is likely for at least Ca^{2+} and Cl^-), one may predict regulation at the level of these cells. Cell parameters that may vary in dependence of the ambient medium are density, spatial distribution, size of apical exposure area, enzymatic make-up of the chloride cells and the permeability characteristics of the junctions between these cells and the surrounding pavement cells.

To examine MR-cell density in filaments of adult fish with the CLSM, individual filaments were incubated in Ringer with the fluorescent probes Con-A and DASPMI (for details see van der Heijden *et al.*, in press). The *in vitro* preparation and labelling procedure of the filaments takes about one hour. Because the filaments were not fixed and kept at a different salinity during labeling (saline instead of freshwater or seawater), checks had to be run to assure that measurements reflected the *in vivo* situation. It appeared that cell density determined in juveniles after *in vivo* labelling was not different from that after *in vitro* labelling.

The results of the dual-labelling show that the branchial epithelium of adult fish consists of multiple layers of DASPMI-positive cells. They are mainly found on the trailing edge and the interlamellar regions of the primary filament epithelium. The epithelium on the trailing edge of tilapia is rather flat and not implanted with secondary lamellae. This facilitates imaging and quantification of MR-cell density and size as well as apical crypt density of a relatively large area of the filament. Only in the toplayer the labelled Con-A was found associated with the apical side of DASPMI-positive cells. We restricted our measurements to this toplayer. Figure 7.3 shows confocal laser scanning micrographs taken from the toplayer of gill filaments from freshwater (Fig. 7.3A) and seawater (Fig. 7.3B) adapted fish, after co-labelling with DASPMI and Con-A. The total number of MR-cells in the toplayer (all the DASPMI-positive cells) were counted, as well as the number of mature cells (double-labelled). Cells that were labelled by DASPMI *only* were considered either accessory, immature or degenerating MR-cells. Cells or particles that were Con-A-positive only, were rarely seen and not included in the countings.

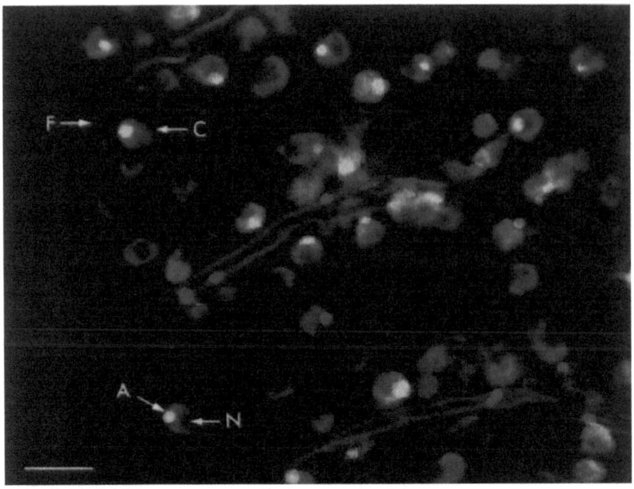

Fig. 7.1

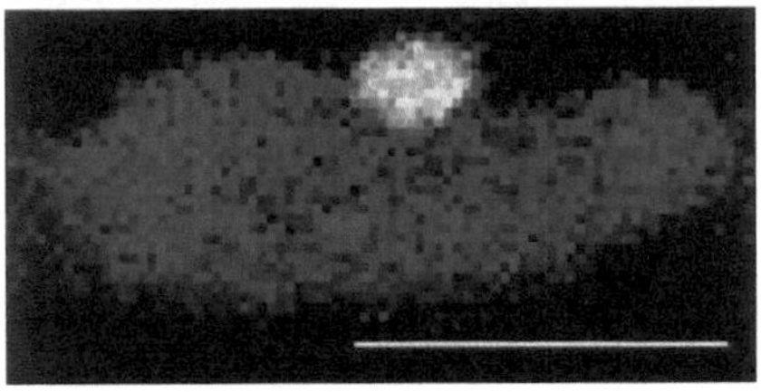

Fig. 7.2

Fig. 7.3a,b

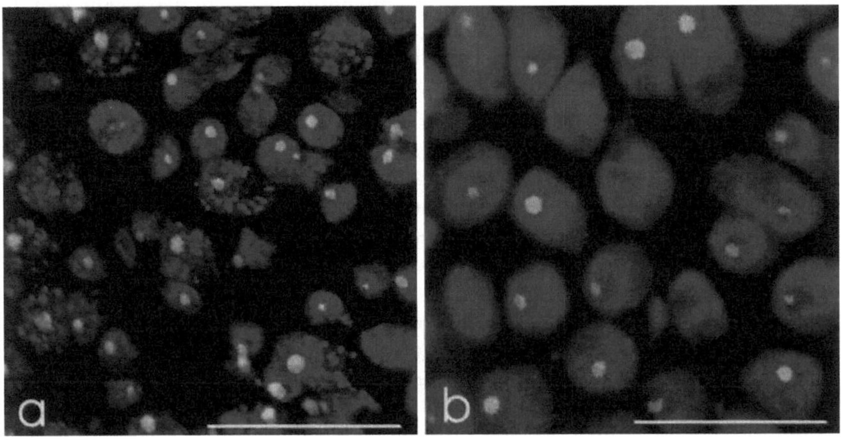

Fig. 7.1. Confocal laser scanning micrograph of the gill filament of a 3-day-old freshwater tilapia after *in vivo* co-labelling with fluorescein conjugated Concanavalin-A (red) and DASPMI (green). The optical section is taken through the filament tips (F indicates the edge of the filament), which point to the left of the picture. Rounded DASPMI-positive cells (C) with a nuclear shade (N) are seen; some of the cells have an apical crypt (A; red spot) facing the water (scale bar 50 μm)

Fig. 7.2. Side-view of 30 superimposed optical sections (0.5 μm thick, 1μm apart) showing a double-labelled cell of a gill filament of an adult freshwater adapted tilapia. DASPMI (red) labels the cytosol, and Concanavalin-A fluorescein conjugate (green) binds to the apical surface of the cell (scale bar 10 μm)

Fig. 7.3. Superimposed picture of consecutive confocal laser scanning micrographs (0.5 μm thick, 1 μm apart) of the trailing edge of colabelled gill filaments from adult freshwater (A) and seawater (B) adapted tilapia. Note that the size of the DASPMI-positive cells (red) is larger in freshwater than in seawater. Green spots indicate Con-A labelled apical crypts (scale bar 50 μm).

The total MR-cell density in the toplayer decreased by 51% from 6233 ± 713 in freshwater fish to 3061 ± 296 ($n.mm^{-2}$ ± S.D.) in fish adapted for 4 weeks to seawater (N = 36 per group). From these cells respectively 3458 ± 553 and 2445 ± 194 ($n.mm^{-2}$ ± S.D.) were double-labelled. This implies that 55% of the MR-cells in freshwater fish are mature, and 80% in seawater fish. Previous TEM studies performed in our lab reported that 55% of all the *chloride* cells in fresh water fish were mature (Wendelaar Bonga *et al.*, 1990). This could indicate that the total MR-cell population as seen with CLSM may represent the total chloride cell population. However, in our study only the toplayer of the multilayered gill epithelium was studied and it may well be that densities of sub-toplayer level are different. Another aspect that differed in seawater fish was the size of the cells, measured as the perimeter of DASPMI-positive cells. Cell size was more that doubled; from 87 ± 9 μm^2 in freshwater fish to 217 ± 69 μm^2 in seawater fish (see Fig. 7.3). A third observation concerned the occurrence of so-called multicellular complexes where two or more MR-cells share one crypt, which is in line with TEM observations by Sardet and co-workers (1979) and Wendelaar Bonga and colleagues (1989). These complexes occurred more frequently in seawater fish (12.8%) than in freshwater fish (1.1%).

We conclude that the fluorescent co-labelling technique with DASPMI and fluorescein conjugated Con-A and examination with CLSM provides a fast and valuable method for the assessment of MR-cells in complex gill filaments, both qualitatively (subtyping, size, multicellular complexes) and quantitatively (numerical density).

7.3 Branchial Ion Transport in Freshwater Fish

Very few measurements of intracellular ion concentrations are available in the literature but are of obvious importance in estimating the electrochemical gradients across the apical and basolateral membranes of the ion transporting cells, which in turn influence the mechanisms and rates of ion transport. For measuring intracellular ion concentrations in chloride cells, methods using ion-selective microelectrodes were not appropriate because of the difficulty to identify these cells and the small size of the apical surface area. With X-Ray microanalysis one can analyse intracellular concentrations of elements relevant to branchial ion transport *in situ*. With this technique only total elemental concentrations can be determined and not the ionic fraction of each element. However, the advantage of this technique is that the total concentrations can be determined with a very high resolution.

7.3.1 X-Ray Microanalysis (XRMA)

Atoms of any element, when struck by electron(s) as occurs in an electron miscoscope, emit X-rays that are characteristic of that element. XRMA entails the use of a suitable detector and analyser to identify and quantify the elements present in a specimen from these characteristic X-rays.

XRMA can be carried out in either the SEM or TEM on specimens that have been fixed chemically or by rapid freezing to a temperature below 173K (cryofixation). Full details of the different techniques can be found in a number of books and reviews of the subject (eg. Marshall, 1987; Echlin, 1992). The methodology of the experiments below has been described in detail by Morgan *et al.* (1994). Briefly, freshly dissected individual filaments were mounted on a microscope stub and frozen on a polished copper block partially submerged in liquid nitrogen. This "metal mirror cooling" prevents the formation of ice crystals in the surface of the specimen (van Harreveld *et al.*, 1974). To expose the internal environment of the cells to the electron beam and produce the flat surface required for consistent X-ray measurement, thin, frozen sections were cut off with a cryo-ultramicrotome and discarded. The specimens were subsequently analysed on the low-temperature stage (113K) of a SEM as described by Morgan *et al.* (1994). Pavement (PE) cells and chloride (MR) cells were distinguished on the basis of shape, size and location and analysed in the apical region, away from the nucleus and intercellular spaces. The concentration, *C*, of the elements in the PE and MR-cells in $mmol.l^{-1}$ tissue water (mM) was calculated from:

$$C = A/B \times S,$$

where *A* is the ratio of net : background X-ray counts of the element in the cell; *B* is the ratio of net : background X-ray counts of the element in the standard and *S* is the concentration of the element in the standard (mM). The relationship between *B* and *S* was linear (Morgan *et al.*, 1994).

The total elemental concentrations obtained by XRMA are the only measurements available of the intracellular composition of individual PE and MR cells of the freshwater fish gill epithelium *in situ*. Other studies that have measured gill intracellular ion concentrations have used either gill scrapings that consist of a mixture of cell types (Munger *et al.*, 1991; Wood and LeMoigne, 1991), or have used preparations that were bathed in a physiological saline such that the cells have not been exposed to true *in vivo* conditions (Pärt and Bergström, 1995). In the next paragraphs we will show how we used XRMA to determine the cellular location and aspects of the control of ion transport in the gills.

7.3.2 Intracellular Elemental Levels of Branchial Epithelial Cells

In Table 7.1 the total concentrations (free and bound forms) of Na, P, Cl and K in PE and MR cells of the brown trout gill are compared to the results of others on the ion transporting cells of the freshwater frog skin (*Rana esculenta*) and on trout gills. The frog skin performs a ionoregulatory function in freshwater similar to the fish gill and contains both MR-cells and cells analogous to pavement cells, known as principal cells (Rick *et al.*, 1978). The intracellular total elemental concentrations of the fish gill cells show reasonable agreement between the two XRMA studies, although there is a difference in the measurements of intracellular Na. Other methods for determination of intracellular elemental levels in fish gills have been done *in vitro*. Munger and co-workers (1991) calculated intracellular total elemental concentrations from estimations of extracellular and intracellular fluid volumes (resp. ECFV and ICFV) with a radiolabelled marker (^{3}H-PEG) and elemental concentrations in the tissue homogenate and blood plasma. Pärt and Bergström (1995) analysed the elemental concentrations of cultured gill cells after permeabilization of these cells in a Na^+ or Cl^--free medium. There is reasonable agreement between all of these measurements.

Recently, ion-selective fluorescent probes, for use in fluorescence microscopy including CLSM, have become available that enable the measurement of free intracellular ion concentrations of cells *in vivo*. These probes have already been used in a variety of vertebrate cell types. The results with sodium probes for instance, indicate that free intracellular ion concentrations are much lower than total concentrations. For fish the first results with these techniques indicate that the same probably holds true for gill cells: ionic Na^+ levels are in the same range as in the cells of higher vertebrates (e.g. Ahlemeyer *et al.*, 1992; Törnquist and Ekokoski, 1993): around 10 mM both in MR and PE cells (J. Li, personal communication), which is 1/6 to 1/3 of the total Na content of gill cells. This value is in line with the $K_{0.5}$ of Na^+/K^+-ATPase for Na^+ of 9,9 mM in tilapia gills (see next chapter by G. Flik) and of 13 mM in trout gills (Flik *et al.*, 1996*a*).

Table 7.1. Intracellular total concentrations ($mmol.l^{-1}$) of Na, Cl, K and P in epithelial cells of freshwater animals.

Brown trout gill PE cell (Morgan *et al.*, 1994)			Frog skin principal cell (Harvey and Kernan, 1984)		
$[Na^+]_{ext}$ (mM)	$[Na^+]_{int}$ (mM)	E_{ap} (estimated)	$[Na^+]_{ext}$ (mM)	$[Na^+]_{int}$ (mM)	E_{ap} (estimated)
0.3	66	-133.8	0.1	2.0	-101
1.0	120	-118.7	1.0	4.0	-87

7.3.3 Cellular Location of Na^+ and Cl^- Uptake

Measurement of the intracellular composition of the PE and MR cells of the brown trout gill epithelium in response to experimental treatments was used to examine one of the major debates of ionic regulation in freshwater teleosts, namely the cellular location of Na^+- and Cl^--uptake mechanisms. The majority of studies have favoured the MR cells as the sites of both Na^+- and Cl^--uptake in fresh water fish, on the basis of the correlation between their distribution and gill ion uptake (e.g. Perry and Laurent, 1989). Few studies however have examined changes in PE cell morphology. Goss *et al.* (1992) found that Na^+-influx during hypercapnia in the brown bullhead (*Ictalurus nebulosus*) was correlated with the development of PE cells and not MR cells. Moreover, following the recent suggestion that Na^+-uptake in freshwater fish is driven by a H^+-ATPase (proton pump) in a manner similar to that of the frog skin (Avella and Bornancin, 1989; Lin and Randall, 1991), proton pumps have been localized to PE cells (Laurent *et al.*, 1994; Sullivan *et al.*, 1995).

To examine the cellular location of Na^+ influx in the brown trout (*S. trutta*) the environmental [Na^+] was manipulated (Morgan *et al.*, 1994). Trout were exposed to low (25μM) external Na^+ for one week in artificial freshwater in which the other major ions remained at control concentrations or to low external Na^+ (25μM) for 1 week and then exposed to 1mM Na^+ for 5 min before sacrifice. The rationale of these experiments was that the intracellular concentration of an ion in an epithelial cell results from a balance between apical and basolateral ion transport. Severe manipulation of the apical influx might be expected to result in a disruption of the normal intracellular concentration of that ion before compensatory mechanisms can be effected. We assume that this disruption would only be seen in the cell type(s) that transport the ion in question. Following exposure to 25μM Na^+, the concentration of Na in the PE cells was slightly (but not significantly) lower than that of the controls (Fig. 7.4A). However, in the fish subsequently exposed to 1mM Na^+, the intracellular Na concentration was almost twice that of controls at approximately 120 mM. Low-Na^+ did not affect Cl concentrations but fish exposed to high-Na^+ had significantly lower concentrations of Cl than controls. In MR cells, the only effect of manipulation of external [Na^+] was a reduction in Cl concentration following exposure to 25μM Na^+ (Fig. 7.4B). After manipulation of the external [Cl^-] concentration in the same way, the intracellular Cl concentrations were significantly increased in both PE and MR cells, with the greater response in the MR cells (Morgan *et al.*, 1994). Exposure of the fish for 1 h to 150μM external thiocyanate (SCN^-), which is a powerful inhibitor of active Cl^- uptake in fish (Kerstetter and Kirschner, 1974, de Renzis 1975), had no effect on the concentration of Na or Cl in PE cells, but resulted in a significant decrease in Cl concentration of the MR cells (Morgan and Potts, 1995*a*). The intracellular Na concentration in MR cells decreased by a similar amount. The correlation between changes in external [Na^+] and intracellular total Na concentration in the PE cells was interpreted therefore as indicating that PE cells and not MR cells are the sites of Na^+-uptake in freshwater fish. Similarly, the decrease in intracellular total Cl in MR cells only, following the exposure of brown trout to 150μM SCN, strongly suggests that only the MR cells are involved in the uptake of Cl^-.

7.3.4 Control and Mechanisms of Branchial Ion Transport

Catecholamines have an effect on branchial ionoregulation primarily as a mechanism for adjusting the acid-base balance of the fish during hypoxia, exercise and stress conditions (Perry and Laurent, 1993). Whether this adjustment is a primary effect on the ion exchange mechanisms or an indirect effect, due to changes in the gill circulation and/or ventilation, is not yet clear. Adrenaline has both α- and β-adrenergic effects, but these can be studied separately using synthetic drugs, that stimulate one type of receptor only. We studied the effects of the β-adrenergic agonist isoproterenol on the intracellular composition of the PE and MR cells in brown trout (Morgan and Potts, 1995*b*). The effects of isoproterenol injection were very similar to those seen in fish exposed to high external Na^+ - there was a significant increase in intracellular Na and Cl. MR cells showed a significant decrease in the Cl concentration only.

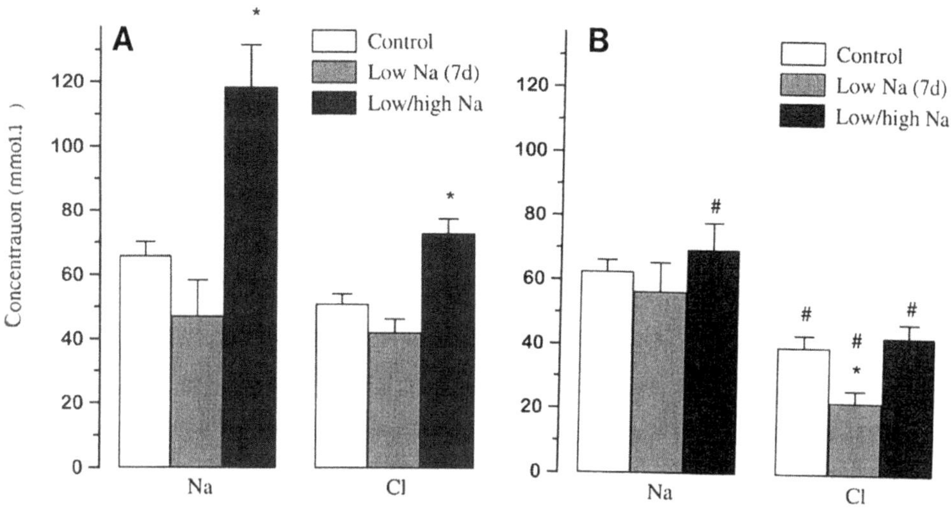

Fig. 7.4. XRMA of the Na and Cl content of (A) pavement (PE) cells and (B) mitochondria-rich (MR) cells of brown trout (*Salmo trutta*) maintained in atrificial fresh waters of varying [Na^+]. * Indicates a value significantly different from the control, # indicates a value in MR cells significantly different from that in PE cells (P<0.05). Values are means ± S.E.M., *N*=30-40. From Morgan *et al.* (1994).

These results suggest that adrenergic agonists, and by implication catecholamines, have a direct effect on branchial ionoregulation. They are in line with findings that β-adrenergic stimulation promotes Na^+-influx (McDonald *et al.*, 1989) and inhibits Cl^--influx (Perry *et al.*, 1984) and may therefore act to clear internal acidosis. The results are also in line with the theory that in freshwater fish, Na^+-uptake occurs via PE cells whilst Cl^--uptake occurs via MR cells.

Perhaps the greatest debate surrounding branchial Na^+- and Cl^-- uptake in freshwater fish concerns the precise mechanisms involved. The uptake of Na^+ has classically been attributed to a 1:1 exchange (antiport) with H^+ and/or NH_4^+ ions, but this mechanism is unlikely to be functional (reviewed by Lin and Randall, 1995). Avella and Bornancin (1989) suggested that Na^+-uptake in freshwater fish occurs through Na^+-channels, driven by an (inside-negative) apical potential generated by a H^+-ATPase (proton pump) as occurs across the freshwater frog skin (Ehrenfeld *et al.*, 1985). Other studies have supported this suggestion (Sullivan *et al.*, 1995; Lin and Randall, 1995).

The availability of XRMA data presents the possibility of estimating the Nernstian potential (driving force) for various ions across the apical membrane of the ion-transporting cells, and comparing these to calculated Nernstian potentials in other epithelia in which ion concentrations have been measured with microelectrodes. However, whilst this is an interesting exercise, the potentials obtained are fundamentally flawed in that XRMA measures total concentration whilst true potentials are calculated from ionic activities which are likely to be significantly lower (see section 7.3.2). For example the Nernstian potential for

Na^+ calculated from the XRMA data across PE cells of the fish gill under normal conditions ($[Na^+]_o$= 0.3 mM and $[Na^+]_i$= 66 mM) is -134 mV. This is much greater than either the calculated Nernstian potential (-78 mV) or the measured apical membrane potential (-101 mV) across the frog skin exposed to 0.1 mM Na^+ (Harvey and Kernan, 1984). Therefore, on the basis of XRMA data, it is difficult to judge if the balance between apical membrane potential and driving force for Na^+ would allow for a passive influx down a Na^+-channel.

Figure 7.5 illustrates three alternative models of Na^+-uptake, all based on the H^+-pump/Na^+ channel mechanism but at different cellular locations. Avella and Bornancin (1989) suggested that the transport mechanism was located on MR cells (Fig. 7.5 I a). A second model (Fig. 7.5 I b) is based on the work of Laurent *et al.* (1994) and Sullivan *et al.* (1995), which localized proton pumps to PE cells in rainbow trout, and concluded that these were the sites of Na^+ uptake. However, this would make the freshwater fish gill the only one out of a number of similar, tight epithelia such as the turtle urinary bladder (Arruda *et al.*, 1990) and mammalian renal collecting tubule (Brown *et al.*, 1988), where the proton pump is not located in MR cells, and does not explain why MR cells proliferate in low Na^+ conditions. In the frog skin model (Fig. 7.5 II), as proposed by Ehrenfeld *et al.* (1985), the Na^+ channels are located on the equivalent of trout PE cells, the principal cells, but the proton pumps occur on the apical membrane of MR cells. The XRMA data presented above are consistent with the models depicted in either Fig. 7.5 I b or 7.5 II. However, none of the three alternatives are consistent with all the available experimental data and thus it seems that further studies are required. Compared to Na^+, Cl^--uptake in freshwater fish is less well understood. Most studies on freshwater fish favour electroneutral Cl^-/HCO_3^- exchange located on MR-cells (e.g. Marshall, 1995) and indeed, the XRMA data presented above provided no evidence to refute this theory.

7.4 Perspectives

CLSM and XRMA offer new possibilities for the study of gill cells in situ. CLSM allows a fast, non-invasive, three dimensional analysis of unfixed gill filaments with site-specific probes as shown in this paper. Apart from the probes used here, that allow quantitation of chloride cells, a wide array of probes is available which allow the study of other aspects of cell physiology. For example, probes for cell proliferation, apoptosis, and necrosis should allow the assessment of cell turnover in the gills in response to certain inhibitors, toxicants or water quality. The use of probes for intracellular pH, Ca^{2+}, Na^+ etc., allow us to study the ion transport in cells.

XRMA allows for the measurement of total intracellular element levels. It is especially useful when experimental manipulations are used to induce shifts in element pool size or composition, which are indicative of both the cellular location and mechanisms of branchial ion transport.

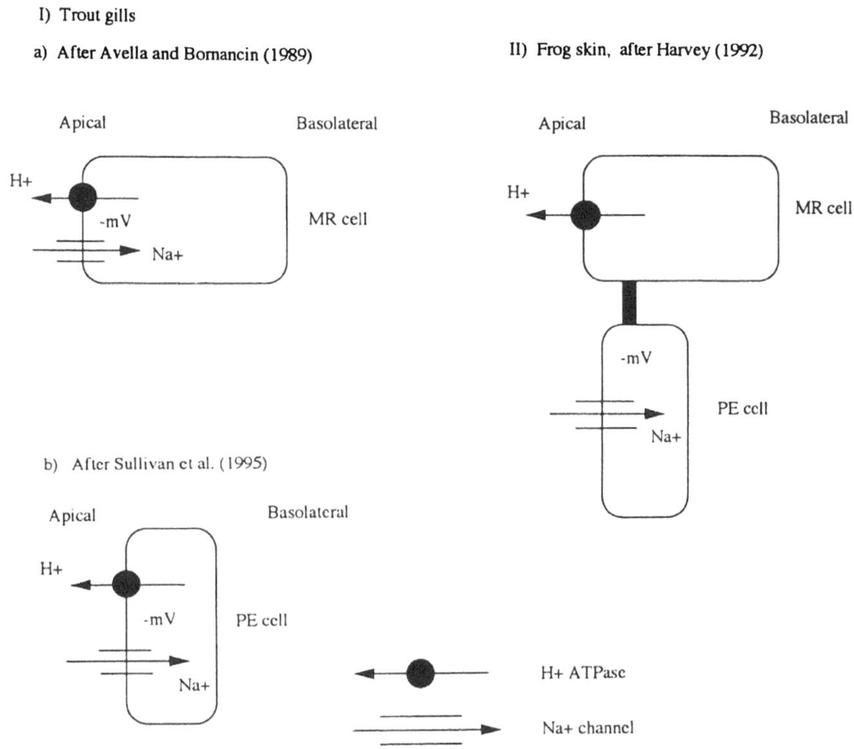

Fig. 7.5 Alternative models of Na^+uptake across (I) the freshwater trout gill and (II) the freshwater frog skin, based on the H^+-pump/Na^+ channel mechanism (see text for an explanation).

References

Alderdice, D.F. (1988). Osmotic and ionic regulation in teleost eggs and larvae. In *Fish Physiology*, vol 11a (ed. W.S. Hoar and D.J. Randall), pp. 163-251. New York: Academic Press.

Arruda, J.A.L., Dytko, G. and Talor, Z. (1990). Stimulation of H^+ secretion by CO_2 in turtle bladder: role of intracellular pH, exocytosis, and calcium. *Am J. Physiol.* 258, R222-231.

Avella, M. and Bornancin, M. (1989). A new analysis of ammonia and sodium transport through the gills of the freshwater rainbow trout (*Salmo gairdneri*). *J. Exp. Biol.* 142, 155-175.

Avella, M., Masoni, A., Bornancin, M. and Mayer-Gostan, N. (1987). Gill morphology and sodium influx in the rainbow trout (*Salmo gairdneri*) acclimated to artificial freshwater environments. *J. Exp. Zool.* 241, 159-169.

Ayson, F.G., Kaneko, T. Hasegawa, S. and Hirano, T. (1994). Development of mitochondrion-rich cells in the yolk-sac membrane of embryos and larvae in tilapia, *Oreochromis mossambicus*, in freshwater and seawater. *J. Exp. Zool.* 270, 129-135.

Bereiter-Hahn, J. (1976). Dimethylaminostyrylmethylpyridiniumiodine (DASPMI) as a fluorescent probe for mitochondria *in situ*. *Biochim. Biophys. Acta* 423, 1-14

Berridge, M. J. and Oschman, J. L. (1972). *Transporting epithelia*. New York London: Academic Press

Brown, D., Roth, J. and Orci, L. (1985). Lectin-gold cytochemistry reveals intercalated cell heterogeneity along rat kidney collecting ducts. *Am. J. Physiol.* 248, C328-C356

Brown, D., Hirsch, S. and Gluck, S. (1988). Localization of a proton-pumping ATPase in rat kidney. *J. Clin. Invest.* 82, 2114-2126.

Coleman, R., Yaron, Z., and Ilan, Z. (1977). An ultrastructural study of the mitochondria-rich cells from the gills of freshwater and seawater adapted *Tilapia aurea* subjected to a pesticide. *J. Fish Biol.*, 11, 589-594

Dagdeviren, A., Alp, H. and Ors, U. (1994). New applications for the zinc iodide-osmium tetroxide technique. *J. Anat.* 184, 83-91.

Echlin, P. (1992). *Low Temperature Microscopy and Analysis*. New York and London, Plenum Press. 539pp.

Ehrenfeld, J., Garcia-Romeu, F. and Harvey, B.J. (1985). Electrogenic active proton pump in *Rana esculenta* skin and its role in sodium ion transport. *J. Physiol., Lond.* 359, 331-355.

Flik, G., Rijs, J.H. and Wendelaar Bonga, S.E. (1985). Evidence for high-affinity Ca^{2+}-ATPase activity and ATP-driven Ca^{2+}-transport in membrane preparations of the gill epithelium of the cichlid fish *Oreochromis mossambicus*. *J. Exp. Biol.* 119, 335-347

Flik, G., Kaneko, T., Greco, A.M. Li, J. and Fenwick, J.C. (1996*a*). Sodium dependent ion transporters in trout gills. *Fish Physiol. Biochem.* in press

Flik, G., Klaren, P.H.M., Schoenmakers, Th.J.M., Bijvelds, M.J.C., Verbost, P.M. and Wendelaar Bonga, S.E. (1996*b*). Cellular calcium transport in fish: unique and universal mechanisms. *Physiol. Zool.* 69, 403-417

Foskett, J.K., Logsdon, C.D., Turner, T., Machen, T.E. and Bern, H.A. (1981). Differentiation of the chloride extrusion mechanism during seawater adaptation of a teleost fish, the cichlid *Sarotherodon mossambicus*. *J. Exp. Biol.* 93, 209-224

Foskett, J.K. and Scheffey, C. (1982). The chloride cell: definitive identification as the salt-secretory cell in teleost. *Science* 215, 164-166

Garcia-Romeu, F. and Masoni, A. (1970). Sur la mise en évidence des cellules à chlorure de la branchie des poissons. *Arch. d'Anat. microsc.* 59, 289-294.

Goss, G.G., Laurent, P. and Perry, S.P. (1992). Evidence for a morphological component in acid-base regulation during environmental hypercapnia in the brown bullhead (*Ictalurus nebulosus*). *Cell Tissue. Res.* 268, 539-552.

Harreveld, A. van, Trubatch, J. and Steiner, J. (1974). Rapid freezing and electron microscopy for the arrest of physiological processes. *J. Microsc.* 100, 189-198.

Harvey, B.J. (1992). Energization of sodium absorption by the H^+-ATPase pump in mitochondria-rich cells of frog skin. *J. Exp. Biol.* 172, 289-309.

Harvey, B.J. and Kernan, R.P. (1984). Intracellular ionic activities in relation to external sodium and the effects of amiloride and/or ouabain. *J.Physiol., Lond.* 349, 501-517

Heijden, A.J.H. van der, Verbost, P.M., Eygensteyn, J., Li, J., Wendelaar Bonga, S.E. and Flik, G.. Mitochondrion-rich cells in gills of tilapia (*Oreochromis mossambicus*) adapted to fresh water or seawater; quantification by confocal laser scanning microscopy. *J. Exp. Biol.*, in press

Hootman, S.R. and Philpott, C.W. (1980). Accessory cells in teleost branchial epithelium. *Am. J. Physiol.* 238, R199-R206.

Hwang, P.P., Tsai, Y.N. and Tung, Y.C. (1994). Calcium balance in embryos and larvae of the freshwater-adapted teleost, *Oreochromis mossambicus*. *Fish Physiol. Biochem.* 13, 325-333.

Karnaky, K.J., Jr., Kinter, L.B., Kinter, W.B. and Stirling, C.E. (1976). Teleost chloride cell. II. Autoradiographic localization of gill Na,K-ATPase in killifish *Fundulus heteroclitus* adapted to low and high salinity environments. *J. Cell Biol.* 70, 157-177

Kerstetter, T.H. and Kirshner, L.B. (1974). HCO_3^--dependent ATPase activity in the gills of the rainbow trout (*Salmo gairdneri*). *Comp. Biochem. Physiol.*48B, 581-589.

Laurent, P.L. (1984). Gill internal morphology. In *Fish Physiology*, vol. Xa (ed. W.S. Hoar and D.J. Randall), pp.73-183. New York: Academic Press.

Laurent, P., Goss, G.G. and Perry, S.F. (1994). Proton pumps in fish gill pavement cells? *Arch. Int. Physiol. Biochim. Biophys.* 102, 77-79.

Li, J., Eygensteyn, J., Lock, R.A.C., Verbost, P.M., Heijden, A.J.H. van der, Wendelaar Bonga, S.E. and Flik, G. (1995). Branchial chloride cells in larvae and juveniles of freshwater tilapia *Oreochromis mossambicus*. *J. Exp. Biol.* 198, 2177-2184

Lin, H. and Randall, D. (1991). Evidence for the presence of an electrogenic proton pump on the trout gill epithelium. *J. Exp. Biol.* 161, 119-134.

Lin, H. and Randall, D.J. (1995). Proton pumps in fish gills In *Cellular and Molecular Approaches to Fish Ionic Regulation* (ed. C.M. Wood and T.J. Shuttleworth), pp. 229-255. San Diego, Academic Press.

Maetz, J. (1974). Aspects of adaptation to hypo-osmotic anf hyperosmotic environments. In *Biochemical and biophysical perspectives in marine biology.* (ed. D.C. Malins and J.R. Sargent) Vol. 1, pp. 1-166. Academic Press, New York.

Marshall, A.T. (1987). Scanning electron microscopy and x-ray microanalysis of frozen-hydrated bulk samples. In *Cryotechniques in biological electron microscopy* (ed. R.A. Steinbrecht and K. Zierold). pp. 240-257. Springer-Verlag, Berlin, Heidelburg.

Marshall, W.S. (1995). Transport processes in isolated teleost epithelia: opercular epithelium and urinary bladder. In *Cellular and Molecular Approaches to Fish Ionic Regulation* (ed. C.M. Wood and T.J. Shuttleworth), pp. 1-23. San Diego, Academic Press.

Marshall, W.S., Bryson, S.E., Midelfart, A. and Hamilton, F.W. (1995). Low conductance anion channel activated by cAMP in teleost Cl^- secreting cells. *Am. J. Physiol.* 268, R963-R969

McCormick, S.D. (1990). Fluorescent labelling of Na^+/K^+-ATPase in intact cells by use of a fluorescent derivative of ouabain: salinity and teleost chloride cells. *Cell Tissue Res.* 260, 529-533.

McCormick, S.D. (1995). Hormonal control of gill Na^+,K^+-ATPase and chloride cell function. In *Cellular and Molecular Approaches to Fish Ionic Regulation* (ed. C.M. Wood and T.J. Shuttleworth), pp. 285-315. San Diego, Academic Press.

McDonald, D.G., Tang, Y. and Boutilier, R.G. (1989). The role of ß-adrenoreceptors in the recovery from exhaustive exercise of freshwater adapted rainbow trout. *J. Exp. Biol.* 147, 471-491.

Moczydlowski, E.G. and Fortes, P.A.G. (1980). Kinetics of cardiac glycoside binding to sodium, potassium adenosine triphosphatase studied with a fluorescent derivative of ouabain. *Biochem.* 19, 969-977.

Morgan, I.J., Potts, W.T.W. and Oates, K. (1994). Intracellular ion concentrations in branchial epithelial cells of brown trout (*Salmo trutta* L.) determined by X-ray microanalysis. *J. Exp. Biol.* 194, 139-151.

Morgan, I.J. and Potts, W.T.W. (1995*a*). The effects of thiocyanate on the intracellular ion concentrations of branchial epithelial cells in brown trout. *J. Exp. Biol.* 198, 1229-1232.

Morgan, I.J. and Potts, W.T.W. (1995*b*). The effects of the adrenoreceptor agonists phenylephrine and isoproterenol on the intracellular ion concentrations of branchial epithelial cells of brown trout (*Salmo trutta* L.). *J. Comp. Physiol.* 165, 458-463.

Munger, R.S., Reid, S.D. and Wood, C.M. (1991). Extracellular fluid volume measurements in tissues of the rainbow trout (*Oncorhynchus mykiss*) *in vivo* and their effects on intracellular pH and ion calculations. *Fish Physiol. Biochem.*, 9(4) 313-323.

Pärt, P. and Bergström, E. (1995). Primary cultures of teleost branchial epithelial cells. In *Cellular and Molecular Approaches to Fish Ionic Regulation* (ed. C.M. Wood and T.J. Shuttleworth) pp207-227. San Diego, Academic Press.

Perry, S.F. and Laurent, P. (1989). Adaptational responses of rainbow trout to lowered external NaCl concentration: contribution of the branchial chloride cell. *J. Exp. Biol.* 147, 147-168.

Perry, S.F. and Laurent, P. (1993). Environmental effects on fish gill structure and function. In *Fish Ecophysiology* (ed. J.C. Rankin and F.B. Jensen). pp. 231-264. London, Chapman and Hall.

Perry, S.F., Payan, P. and Girard, J.P. (1984). Adrenergic control of branchial chloride transport in the isolated perfused head of the freshwater trout (*Salmo gairdneri*). *J. Comp. Physiol.* 154B, 269-274.

Perry, S.F., Goss, G.G. and Fenwick, J.C. (1992*a*). Interrelationships between gill chloride cell morphology and calcium uptake in freshwater teleosts. *Fish Physiol. Biochem.* 10, 327-337

Perry, S.F., Goss, G.G. and Laurent, P. (1992*b*). The interrelationships between gill chloride cell morphology and ionic uptake in four freshwater teleosts. *Can. J. Zool.* 70, 1775-1786

Pisam, M. and Rambourg, A. (1991). Mitochondria-rich cells in the gill epithelium of teleost fishes: an ultrastructural approach. *Int. Rev. Cyt.* 130, 191-232.

Pisam, M., Sardet, C. and Maetz, J. (1980). Polysaccharidic material in chloride cell of teleostean gill: modifications according to salinity. *Am. J. Physiol.* 238, R213-R218

Pisam, M., Caroff A. and Rambourg A. (1987). Two types of chloride cells in the gill epithelium of a freshwater adapted euryhaline fish: *Lebistes reticularis*; their modifications during adaptation to seawater. *Am. J. Anat.* 179, 40-50

Pisam, M., Auperin, B., Prunet, P., Rentier-Delrue, F., Martial, J. and Rambourg, A. (1993). Effects of prolactin on α and ß chloride cells in the gill epithelium of the saltwater adapted tilapia "*Oreochromis niloticus*". *Anat. Rec.*, 235, 275-284

Powell, M. D., Speare, D. J. and Wright, G. M. (1994). Comparative ultrastructural morphology of lamellar epithelial, chloride and mucus cell glycocalyx of the rainbow trout (*Oncorhynchus mykiss*) gill. *J. Fish Biol.* 44, 725-730

Pratap, H. B. and Wendelaar Bonga, S. E. (1993). Effect of ambient and dietary cadmium on pavement cells, chloride cells, and Na^+/K^+-ATPase activity in the gills of the freshwater teleost *Oreochromis mossambicus* at normal and high calcium levels in the ambient water. *Aquat. Toxicol.* 26, 133-150

Renzis, G. de (1975). The branchial chloride pump in the goldfish *Carassius auratus*: rèlationship between Cl^-/HCO_3^- and Cl^-/Cl^- exchanges and the effect of thiocyanate. *J. Exp. Biol.* 63, 587-602.

Rick, R. (1992). Intracellular ion concentrations in the isolated frog skin epithelium: evidence for different types of mitochondria-rich cells. *J. Membr. Biol.* 127, 227-236.

Rick, R., Dörge, A., von Arnim, E. and Thurau, K. (1978). Electron microprobe analysis of frog skin epithelium: evidence for a syncytial sodium transport compartment. *J. Membr. Biol.* 39, 313-331.

Sardet, C., Pisam, M. and Maetz, J. (1979). The surface epithelium of teleostean fish gills. Cellular and juctional adaptations of the chloride cell in relation to salt adaptation. *J. Cell Biol.* 80, 96-117.

Schuster, V.L., Bonsib, S.M. and Jennings, M.L. (1986). Two types of collecting duct mitochondria-rich (intercalated) cells: lectin and band 3 cytochemistry. *Am. J. Physiol.* **251**, C347-C355

Sullivan, G.V., Fryer, J.N. and Perry, S.F. (1995). Immunolcalization of proton pumps (H^+-ATPase) in pavement cells of rainbow trout gill. *J. Exp. Biol.* 198, 2619-2629.

Wendelaar Bonga, S.E. and Meij, J.C.A. van der (1989). Degeneration and death, by apoptosis and necrosis, of the pavement and chloride cells in the gills of the teleost *Oreochromis mossambicus*. *Cell Tissue Res.* 255, 235-243.

Wendelaar Bonga, S.E., Flik, G., Balm, P.H.M. and Meij, J.C.A. van der (1990). The ultrastructure of chloride cells in the gills of the teleost *Oreochromis mossambicus* during exposure to acidified water. *Cell Tissue Res.* 259, 575-585.

Witters, H., Berckmans, P. and Vangenechten, C. (1996). Immunolocalization of Na^+, K^+-ATPase in the gill epithelium of rainbow trout, *Onchorhynchus mykiss*. *Cell. Tiss. Res.* 283, 461-468.

Wood. C.M. and LeMoigne, J. (1991). Intracellular acid-base responses to environmental hyperoxia and normoxic recovery in rainbow trout. *Respir. Physiol.* 86, 91-113.

Zadunaisky, J.A. (1984). The chloride cell: The active transport of chloride and the paracellular pathways. In *Fish Physiology Vol. XB* (ed. W.S. Hoar and D.J. Randall). pp. 129-176. Orlando, Academic Press.

Transport and Housekeeping of Calcium in Fish Gills

G. Flik

Department of Animal Physiology, Faculty of Science,
University of Nijmegen, Toernooiveld, 6525 ED Nijmegen
The Netherlands

8.1 Introduction

The epithelium covering the gills of freshwater fish is considered a tight epithelium with very little ion traffic along paracellular routes. Branchial uptake of ions in fresh water, therefore, requires that ions move through the cells crossing the plasma membrane barrier twice. But in the sea water environment, the branchial epithelium shows characteristics of a leaky epithelium, allowing ion movement(s) through the cells as well as along paracellular routes.

Freshwater fish take up sodium, chloride and calcium from the water, sea water fish extrude sodium and chloride and take up calcium (Marshall, 1995; Lin and Randall, 1995; Flik et al., 1995b). For discussions on current concepts of the transport routes for sodium and chloride in fish gills, see some of the other chapters in this volume. Consensus exists that calcium transport in fish gills is mediated by the chloride cells (also called ionocytes or mitochondrion-rich cells) in the epithelium. Flat epithelia from the buccal cavity of trout (Marshall et al., 1992), tilapia (McCormick et al., 1992) and killifish (Marshall et al., 1995), that contain chloride cells have been used in Ussing chamber set-ups to demonstrate that Ca^{2+} transport by these epithelia is strictly dependent on the presence of chloride cells (cell densities estimated by staining the mitochondrion-rich chloride cells with vital stains for mitochondria; e.g. McCormick, 1990). Similar, yet more circumstantial evidence has been presented for correlation between whole body calcium fluxes - that reflect branchial calcium fluxes - and chloride cell density or surface area of chloride cells in contact with the water in the gills (Perry et al., 1992).

Ca^{2+} taken up from the water enters the chloride cells passively through a Ca^{2+}-conductive pathway that may be blocked by lanthanum (Perry and Flik, 1988; Verbost et al., 1987) and is closed by the hormone stanniocalcin (Verbost et al., 1993). Being hormone-sensitive, a second messenger operated calcium channel has been predicted for the Ca^{2+} entry step, but evidence for this postulate is as yet circumstantial and awaits patch-clamp and molecular biological analyses of chloride cells. As cytosolic Ca^{2+} in isolated chloride cells at rest is around 100 nM (Li et al., 1996) and the resting potential of the cell is estimated at - 60 mV (Marshall, 1995; no direct measurements available in the literature), it follows that the rate of calcium uptake will proceed largely independently from the water calcium content and that regulation of calcium uptake requires primarily a mechanism that can stop the inflow of Ca^{2+} into the epithelial cells. Little is known concerning the buffering and traffic of Ca^{2+} within the cytosol of the chloride cell. However, the chloride cell endoplasmic reticulum has an IP_3-

sensitive channel and a thapsigargin-sensitive Ca^{2+}-ATPase and the mitochondria store calcium (Flik et al., 1993). Further, the basolateral plasma membrane of these cells is well supplied with mechanisms for calcium extrusion. The basolateral plasma membrane forms an elaborate and interconnecting system of tubular invaginations, creating an intracellular analogue of the basal labyrinth observed in e.g. the mammalian nephron. Such membrane extensions are assumed to reflect the need for increased capacity. It is in this tubular system where the branchial epithelial machinery for extrusion of Na^+ to the serosal compartment, the Na^+/K^+-ATPase, is concentrated. In the same membrane reside a Ca^{2+}-ATPase and a Na^+/Ca^{2+}-exchanger that can move calcium to the serosa at the expense of energy stored in ATP or in a Na^+-gradient, respectively (Flik et al., 1995a). The Na^+/K^+-ATPase maintains an inwardly directed sodium gradient across the basolateral plasma membrane of around 140 mM (plasma Na: 150-160 mM; intracellular Na^+: 11 mM), which is sufficient to drive uphill calcium transport under the electrochemical conditions assumed to prevail near the basolateral plasma membrane (cytosolic Ca^{2+}: 100 nM; plasma Ca^{2+}: 1.25 mM; a transmembrane potential difference somewhat smaller than -60 mV, assuming a slightly negative transepithelial potential difference). To fully comprehend active Ca^{2+} transport by chloride cells we must consider the Na^+/K^+-ATPase, the Ca^{2+}-ATPase and the Na^+/Ca^{2+}-exchanger.

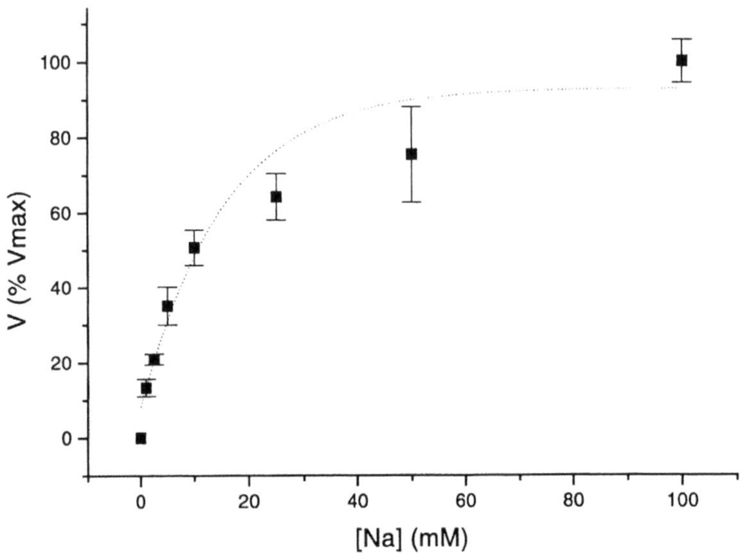

Fig. 8.1. Na^+-dependence of Na^+/K^+-ATPase activity in plasma membranes of freshwater tilapia branchial epithelium. For details on membrane isolation and assay media see Flik et al., 1994. Resealed membranes were permeabilised with saponin (0.2 mg/mg protein) to optimise substrate accessibility. When $[Na^+]$ was varied equimolar replacement with $NMDG^+$ secured isotonicity of the media. Maximum velocity was 52 ± 3.5 μmol P_i/h per mg protein (N=5). Velocities are have been expressed as percentage of V_{max} (= 100%). The calculated halfmaximal activation concentration of Na^+ was 9.9 ± 1.3 mM.

Although it is common practise to determine Na^+/K^+-ATPase in gill homogenates and membrane preparations, a value for the Na^+ affinity of the tilapia enzyme is, to the best of our knowledge, not available in the literature. On the other hand, the pivotal role of this enzyme in ion transport in gills together with its possible regulation by Na^+ requires that this information is available. Therefore a kinetic analysis was performed. The Na^+/K^+-ATPase of tilapia gill is half-maximally activated by 9.9 mM Na^+ (Figure 8.1) , strongly suggesting that the enzyme is regulated by cytosolic Na^+ levels (Ewart and Klip, 1995). A similar affinity for Na^+ (13 ± 6 mM Na^+) was found for Na^+/K^+-ATPase in freshwater trout gills (Flik et al., 1996) as well as for Na^+/K^+-ATPase in tilapia enterocytes (Flik et al., 1990). Assuming then that cytosolic Na^+ performs a pivotal role in the regulation of Na^+/K^+-ATPase activity, it follows that *in vivo* a secondary transporter such as the Na^+/Ca^{2+}-exchanger may be indirectly controlled by Na^+ through its action on Na^+/K^+-ATPase. Also, the activity of the Na^+/K^+-ATPase may be predicted to govern the electrical conditions at the membrane level that determine the reversal of the transport mode of the exchanger (see below).

An important premise for the operation of a Ca^{2+}-ATPase in transport is that the enzyme is activated by prevailing cytosolic Ca^{2+} concentrations and translocates Ca^{2+} across the membrane. The half-maximal activation concentrations of Ca^{2+} for ATP-driven Ca^{2+}-transport in a variety of fish gill plasma membrane preparations is in the nanomolar range (Flik et al., 1995a), making the enzyme subject to regulation by variations in cytosolic Ca^{2+}. Estimates for free calcium levels in chloride cell cytosol come to around 100 nM Ca^{2+}, a value comparable to those reported in other vertebrate cells. Obviously, assays that are based on the translocation of Ca^{2+} into a vesicular space do allow the discrimination of Ca^{2+}-transporting ATPase from e.g. Ca^{2+} stimulated ecto-ATPase that is not involved in transport (Flik and Verbost, 1995).

The operation of the Na^+/Ca^{2+}-exchanger is governed by the electrochemical conditions near the plasma membrane and for this carrier the situation is less clear. Half-maximum activation by Ca^{2+} occurs at micromolar rather than nanomolar concentrations (typical values are 2 - 5 μM Ca^{2+}) and as micromolar concentrations seldom occur in the cell, this would mean that this exchanger operates only at a few percent of its capacity. The half- maximum activation concentration for Na^+ is around 40 mM and this would mean that, for the carrier in its extrusion mode, plasma Na^+ levels (over 100 mM) saturate the carrier and will have no regulatory role. As the carrier activity is governed by the gradients for Ca^{2+} and Na^+ across the membrane, the electrical conditions should of course be considered as well. Proceeding from 158 mM Na^+ and 1.25 mM Ca^{2+} in plasma of a freshwater tilapia, 10.7 mM Na^+ and 95 nM Ca^{2+} in the cytosol of the chloride cell give a reversal potential,

$$E_{Na,Ca} = RT/F \ln \{(Na_o^3 \times Ca_i)/(Na_i^3 \times Ca_o)\}, \qquad (1)$$

of -35.46 mV. This means that the carrier at the predicted, lower potential differences in the cytosol will operate in the extrusion mode; since the reversal potential is close to the predicted potential difference across the cell membrane, the carrier can only be moderately active *in vivo*. This notion adds to the chemical considerations on the predicted low activity of this carrier. Taken together the carrier would be active far below its maximum capacity and its activity rather sensitive to changes in electrical conditions.

Summarising the above it would appear that predictions on the involvement of carriers in Ca^{2+} transport in a fish gill require biochemical, and in particular kinetic data, on the carriers in question preferably in combination with (electro) physiological data on the cell and the intact tissue. This is a particular requirement for the gills where preferably three carriers should be analysed. A somewhat simpler situation occurs in the intestine and kidney of the tilapia: in plasma membranes of enterocytes a powerful Na^+/Ca^{2+}-exchanger appears to be the exclusive Ca^{2+} extrusion mechanism, whereas in renal plasma membranes a Ca^{2+}-ATPase serves this function. Adaptation of tilapia to sea water did not affect Ca^{2+}-ATPase or Na^+/Ca^{2+}-exchanger activity in the gills (as compared to freshwater controls) in line with equal calcium transport rates observed in the gills under the two conditions. However, in the enterocyte the Na^+/Ca^{2+}-exchanger activity and in the renal tissue the Ca^{2+}-ATPase activity decreased in line with a decrease intestinal calcium absorption and a decreased need for Ca^{2+} reabsorption in the kidney, respectively (Flik et al., 1995b).

8.2 Overcapacity of transporters; Ca^{2+} transport in stanniectomized eels

In freshwater eels, removal of the corpuscles of Stannius - the source of the antihypercalcemic hormone stanniocalcin (see Flik et al., 1995a) - invariably leads to a 4- to 5-fold increase in branchial Ca^{2+} inflow (Ca^{2+} inflow values are around 100 nmol/h per 100 g in controls and 400 to 500 nmol/h per 100 g in stanniectomised eels and this increase is attributed to a loss of control over Ca^{2+} entry into the chloride cell (Verbost et al., 1993). As the electrophysiology of the gills is not affected in stanniectomised eels (Verbost et al., 1993), we may assume that the increased flow follows the transcellular route solely, and thus the extrusion machinery of the chloride cells must pump more than 4-fold the amount of calcium per unit of time than normal. Such increases in Ca^{2+} flow must be considered supraphysiological, so one would predict, given the observation that stanniectomised eels will live for months with the resulting dramatic hypercalcemia, that the extrusion machinery is adapted to this Ca stress by increasing the capacity of the carriers involved. However, no changes in branchial plasma membrane ATP-driven Ca^{2+} transport, in Na^+/Ca^{2+}-exchanger activity (AJH van der Heijden, personal communication) or in Na^+/K^+-ATPase activity have been observed (Verbost et al., 1993). Moreover, the increase in plasma Ca^{2+} (from 1.25 to 3.03 mM) at a constant plasma Na and a presumed unaffected cytosolic Na^+ level will shift the equilibrium potential for the exchanger from -40 mV to - 63 mV (according to formula 1). Clearly, the exchanger essentially would be inactivated by this shift in plasma Ca^{2+} and may even operate in the calcium import mode; the conclusion would be that Ca^{2+} transport in gills of stanniectomised fish will not depend on Na^+/Ca^{2+}-exchanger activity. The Ca^{2+}-ATPase on the other hand, must have a significant overcapacity to manage a fourfold larger export of Ca^{2+} to the serosal compartment. The matter of overcapacity of enzyme activity was addressed before (Flik and Verbost, 1993) and a 200-fold overcapacity was calculated for the Ca^{2+}-ATPase in tilapia gills, comparing capacity of the ATP-driven Ca^{2+} pump (based on maximum velocities of the enzyme) with branchial Ca^{2+} flow. We conclude from such calculations that the biochemical assay conditions in all likelihood demonstrate enzyme activity that is silent *in vivo.*

8.3 Involvement of carriers in transepithelial Ca^{2+} transport

8.3.1 Prolactin and Ca^{2+} transport in tilapia

In spite of the clear potential overcapacity of the Ca^{2+}-ATPase in tilapia gills, treatment of this fish with prolactin further enhances the activity of this transporter. The significance of this can be seen when it is noted that the prolactin cells of freshwater tilapia (Nijmegen tap water [Ca] : 0.7 mM) are only moderately active and become rapidly *in*activated when the fish are exposed to "high calcium" water (e.g. sea water). Conversely, the cells are strongly activated when ambient calcium levels are decreased (Wendelaar Bonga et al., 1983). A prolonged (8 days) treatment with prolactin thus induces conditions of enhanced branchial calcium transport similar to that seen in fish kept in "low calcium" (0.2 mM Ca) water (Flik et al., 1986). Indeed, branchial Ca^{2+} inflow increases by up to 60% when tilapia are treated with prolactin. Proceeding from the notion that there may be an inherent 200-fold overcapacity of the ATP-driven Ca^{2+} pump in tilapia gills one would not predict any changes in the related biochemistry. However, the biochemistry of prolactin treated tilapia gills is dramatically altered: Ca^{2+}-ATPase activity increases and Na^{+}/K^{+}-ATPase activity decreases (Flik et al., 1994). Further, in the prolactin-treated fish the opercular epithelium undergoes marked changes in chloride cell density (a dose-dependent effect of prolactin with a threefold maximum increase over control levels). Also, the prolactin treatment is associated with a dose-dependent increase in basal cortisol levels. Cortisol *per se* is known to induce chloride cell proliferation and to enhance Na^{+}/K^{+}-ATPase activity (McCormick, 1990b). These observations indicate that the prolonged prolactin treatment may have induced a new population of Ca^{2+} transporting cells (chloride cells), with an enzymatic composition that reflects the prolactin and cortisol status of the fish, which in turn may determine its adaptability to soft water. The increased Ca^{2+}-ATPase activity that correlates so well with the branchial Ca^{2+} inflow favours the conclusion that the ATP-driven Ca^{2+} pump serves a role in transepithelial Ca^{2+} transport.

8.3.2 Cadmium and Ca^{2+} transport in trout and tilapia

More evidence for a direct dependence of branchial Ca^{2+} inflow on Ca^{2+}-ATPase activity comes from studies on the inhibitory action of waterborne cadmium on calcium uptake from the water (Verbost et al., 1987). For example, in trout Cd^{2+} in the ambient water uses the stanniocalcin-sensitive Ca^{2+} entry pathway to enter the chloride cell where it can eventually inhibit Ca^{2+} uptake. The lag time between the start of the exposure to Cd^{2+} and the inhibition of Ca^{2+} uptake appeared to be shorter with increasing Cd^{2+} concentrations in the water. As the concentration ratio's for Cd over Ca (0.1 - 1 μM Cd, 700 μM Ca) in the water are small it is unlikely that competition between Cd^{2+} and Ca^{2+} for entry underlies the inhibition of Ca^{2+} uptake. However, once the Ca^{2+} extrusion machinery becomes inhibited by Cd, the resulting rise in cytosolic Ca^{2+} may feedback on the apical membrane calcium channel and result in a blockage of Ca^{2+} entry (Verbost et al., 1989). That increases in intracellular Ca^{2+} may inhibit Ca^{2+} entry into chloride cells was also shown for Fundulus opercular chloride cells (Marshall et al., 1995). Additionally, the Ca^{2+}-ATPase in the basolateral plasma membrane was found to be an extremely sensitive target for Cd^{2+} (e.g. 50% inhibition of ATP-driven Ca^{2+} transport at 250 nM Ca^{2+} by 3 nM Cd^{2+}; Schoenmakers et al., 1992). In a tilapia enterocyte basolateral plasma membrane preparation it was demonstrated (Schoenmakers et al., 1992) that the Ca^{2+}-ATPase mediated Ca^{2+} transport has a K_i-value for Cd^{2+} of 3 pM, the Na^+/Ca^{2+}-exchanger mediated transport of Ca^{2+} a K_i of 11 nM. As Cd^{2+} competes with Ca^{2+} for binding to the transport sites for Ca^{2+} on the carriers (Verbost et al., 1989; Schoenmakers et al., 1992), we attribute the difference in Cd^{2+}-sensitivity to the apparent differences in affinity for Ca^{2+} of the carriers. The conclusion of these experiments then is that branchial calcium uptake is inhibited primarily by a specific blockade of the Ca^{2+}-ATPase in the basolateral plasma membrane and that Cd^{2+} thus serves as a probe to demonstrate the involvement of Ca^{2+}-ATPase mediated transport of Ca^{2+}. Interestingly, Cd is transported across the gills into the blood compartment and, although no evidence is available for branchial plasma membranes, it is tempting to speculate that at least part of the Cd transport across the plasma membrane is mediated by the Na^+/Ca^{2+}-exchanger therein: in a tilapia enterocyte plasma membrane preparation it was demonstrated that the exchanger in a Ca^{2+}/Ca^{2+}-exchange mode may exchange Cd^{2+} for Ca^{2+} and thus facilitate cadmium traffic across the membrane at the rates observed that cannot be explained on the basis of simple diffusion.

8.3.3 Na^+/Ca^{2+}-exchanger activity and transepithelial transport

Very few studies have addressed the involvement of Na^+/Ca^{2+}-exchanger activity in chloride cell mediated Ca^{2+} transport. In the chloride cell-rich opercular membrane of the euryhaline *Fundulus* mounted in an Ussing chamber set-up it was shown that at least a part of the Ca^{2+} influx across this epithelium is Na^+-dependent (Verbost et al., 1996). Here an analogy is seen with Ca^{2+} transport across tilapia stripped intestinal epithelium (Flik et al., 1990), that is Na^+-dependent and in all likelihood reflecting a Na^+/Ca^{2+}-exchanger-mediated transport; Ca^{2+}-ATPase activity in the tilapia enterocyte is negligible. However, when testing the hypothesis that the *Fundulus* opercular epithelium from fish adapted to sea water would be more dependent on Na^+/Ca^{2+}-exchange activity for transepithelial Ca^{2+} transport - assuming more Na^+ traffic under these conditions - it appeared that relatively less Ca^{2+} transport could be correlated with Na^+-dependent mechanisms (Verbost et al., 1996). This study thus shows that Na^+/Ca^{2+}-exchanger activity may be may be important for a limited part of the transport mediated by chloride cells.

An often heard statement is that the Na^+/Ca^{2+}-exchanger provides the cells of Ca^{2+}-transporting epithelia with an emergency mechanism for extrusion of Ca^{2+}

(in contrast to e.g. heart cells or neurones in which the exchanger is pivotal in sodium and calcium handling) and plays a minor role in Victoria calcium transport. This statement stems from the prediction that the activity of this carrier is low, based on the kinetic parameters determined *in vitro,* and the narrow range between the presumed potential difference across the plasma membrane and the reversal potential of the carrier. Yet, our understanding of electrochemical conditions at the subcellular level may be too limited for a proper interpretation of the activity of this carrier when it appears that a significant part of branchial Ca^{2+} transport is Na^+-dependent and thus Na^+/Ca^{2+}-exchanger mediated?

8.4 Housekeeping

There is no evidence available to attribute an exclusive role to the Ca^{2+}-ATPase or to the Na^+/Ca^{2+}-exchanger in either transepithelial transport or housekeeping (cell homeostasis related activity). Nevertheless, the presence of two carriers for Ca^{2+} with such widely different kinetic properties and distribution gives rise to and maintains this concept. One could argue of course that a subdivision of tasks (extrusion for transport or for homeostasis) is purely artificial as proper extrusion for transport should not differ qualitatively from extrusion for homeostasis. A quantitative upregulation of carrier density correlated with enhanced transport is evidence, albeit still circumstantial, for an involvement of a carrier in vectorial transport. One could try to find then a situation where more stress is put on the housekeeping activity of the calcium carriers in the gills. Such a situation may be presented by keeping euryhaline fish in sea water where it is assumed that the turnover of sodium and calcium in the gill tissue will be much higher than in fresh water (Maetz, 1974). We have carried out such experiments with tilapia and trout, fish that show interesting and different responses to this treatment (Table 8.1).

Table 8.1. Na^+/K^+-ATPase, Ca^{2+}-ATPase and Na^+/Ca^{2+}-exchange in plasma membranes of gills of freshwater (FW) and sea water (SW) tilapia and trout

	Tilapia	Tilapia	Trout	Trout
	FW	SW	FW	SW
Na^+/K^+-ATPase	630	922	113	388*
Ca^{2+}-ATPase	24.5	30.2	2.4	18.9*
Na^+/Ca^{2+}-exchange	25.1	30.8	4.9	24.3*

Enzyme activities were converted to nmol/min per mg membrane protein and corrected for % inside-out oriented vesicles (Ca^{2+}-ATPase) or % resealing (Na^+/Ca^{2+}-exchange, assumed to operate in all resealed membranes); Na^+/K^+-ATPase activity (sodium- and potassium-dependent, ouabain-sensitive ATP-hydrolytic activity) was determined in membranes treated with saponin to unmask activity that is latent as a result of membrane resealing. * significantly different from FW group. Data taken and adapted from Verbost et al., 1994, and from Flik et al., 1996.

Whereas in fresh water and sea water tilapia gills no significant differences in Na^+/K^+-ATPase, Ca^{2+}-ATPase or Na^+/Ca^{2+}-exchanger activities were found, the activities of all three carriers are upregulated in the trout gills. Unfortunately, we have no data on calcium transport in sea water trout. Nevertheless, the outcome of these experiments could be interpreted at least in two ways. First, it could be that the Ca^{2+} uptake via the gills in sea water trout is larger then in fresh water and that the increased carrier activities relates to this. Second, the trout branchial epithelium may be more permeable to Na^+ and Ca^{2+} as compared to tilapia. In that case, the increased carrier capacity in the sea water trout could reflect a requirement for cell homeostatic purposes. Clearly more research is needed in this field.

There is, however, a carrier that may be assigned a specific role in either transport or housekeeping. Trout gills display a bumetanide-sensitive $Na^+/K^+/Cl^-$-cotransporter activity, that is upregulated after transfer of the fish to sea water (Flik et al., 1996). In fresh water the cotransporter can only be involved in cell volume regulation as freshwater trout take up chloride in exchange for bicarbonate (energised uptake) at the apical membrane followed by passive movement of chloride through basolateral Cl^- channels, and thus do not require the cotransporter for chloride transport; in sea water, however, chloride cells extrude Cl^- and to this end Cl^- is imported from the serosa into the cell along a bumetanide-sensitive pathway and subsequently secreted to the water down the electrochemical gradient through a CFTR-like channel in the apical membrane of the chloride cell (Marshall, 1995). The demonstration of a bumetanide-sensitive cotransporter in both fresh water and sea water gill plasma membrane preparations would favour a housekeeping function, viz. volume regulation in fresh water, and a transport function, viz. for Cl^- extrusion, in sea water gills.

Future biochemical studies on carrier proteins involved in transepithelial transport may benefit from parallel physiological or endocrinological studies that allow to draw conclusions on the direction of transport or give predictable changes in transport activity.

References

Ewart, H.S. and Klip, A (1995) Hormonal regulation of the Na^+-K^+-ATPase: mechanisms underlying rapid and sustained changes in pump activity. Am. J. Physiol. 269: C295-C311

Flik, G, Fenwick, JC, Kolar, Z, Mayer-Gostan, N, and Wendelaar Bonga, SE (1986) Effects of low ambient calcium levels on whole body Ca^{2+} flux rates and internal calcium pools in the freshwater cichlid teleost, *Oreochromis mossambicus*. J. Exp. Biol. 120: 249-264.

Flik, G, Schoenmakers, ThJM, Groot, JA, van Os, CH, and Wendelaar Bonga, SE (1990) Calcium absorption by fish intestine: the involvement of ATP- and sodium-dependent calcium extrusion mechanisms. J. Membrane Biol. 113: 13-22.

Flik, G, Van der Velden, JA, Dechering, KJ, Verbost, PM, Schoenmakers, ThJM, Kolar, ZI and Wendelaar Bonga, SE (1993) Ca^{2+} and Mg^{2+} transport in gills and gut of tilapia, *Oreochromis mossambicus*: a review. J. Exp. Zool. 265, 356-365.

Flik, G and Verbost PM (1993) Calcium transport in fish gills and intestine. J. exp. Biol. 184, 17-29.

Flik, G, Rentier-Delrue, F and Wendelaar Bonga, SE (1994a) Calcitropic effects of recombinant prolactins. Am. J. Physiol. 266, R1302-R1308.

Flik, G and Verbost, PM (1995) Cellular mechanisms in calcium transport and homeostasis in fish. In: Biochemistry and molecular biology of fishes; Hochachka, PW and Mommsen, TP eds., pp 251-265, Elsevier, Amsterdam.

Flik, G, Verbost, PM and Wendelaar Bonga, SE (1995a) Calcium transport processes in fishes. In: Fish Physiology, Vol. XIV, cellular and molecular approaches to fish ionic regulation; Wood, CM and Mommsen, TP eds., pp 317-342. Academic Press, San Diego.

Flik, G, Klaren, PHM, Schoenmakers, ThJM, Bijvelds, MJC, Verbost, PM and Wendelaar Bonga, SE (1995b) Cellular calcium transport in fish: unique and universal mechanisms. Physiol. Zool. 69: 403 - 417.

Flik, G, Kaneko, T, Greco, AM, Li, J and Fenwick, JC (1996) Sodium dependent ion transporters in trout gills. Fish Physiol. Biochem.: In press.

Li, J, Lock, RAC, Wendelaar Bonga, SE, and Flik, G (1996) Some aspects of NA^+ and Ca^{2+} homeostasis in isolated chloride cells of the teleost *Oreochromis mossambicus*: a confocal laserscanning microscopical analysis. J. Exp. Biol.: submitted.

Lin, H and Randall, DJ (1995) Proton pumps in fish gills. In: Fish Physiology, Vol. XIV, cellular and molecular approaches to fish ionic regulation; Wood, CM and Mommsen, TP eds., pp 229-257, Academic Press, San Diego.

Maetz, J. (1974) Aspects of adaptation to hypo-osmotic and hyperosmotic environments. In: Biochemical and Biophysical Perspectives in Marine Biology, Vol. I; Malins, DC and Sargent, JR eds., pp 1-167, Academic Press, New York

Marshall, WS, Bryson, SE and Wood, CM (1992) Calcium transport by isolated skin of rainbow trout. J. Exp. Biol. 166: 297-316.

Marshall, WS, Bryson, SE, Burghardt, JS and Verbost, PM (1995) Ca^{2+} transport by opercular epithelium of the fresh water adapted euryhaline teleost, *Fundulus heteroclitus*. J. Comp. Physiol. B 165: 268-277.

Marshall, WS (1995) Transport processes in isolated teleost epithelia: opercular epithelium and urinary bladder. In: Fish Physiology, Vol. XIV, cellular and molecular approaches to fish ionic regulation; Wood, CM and Mommsen, TP eds., pp 1-25, Academic Press, San Diego.

McCormick, SD (1990a) Fluorescent labelling of Na^+, K^+-ATPase in intact cells by use of a fluorescent derivative of ouabain: Salinity and teleost chloride cells. Cell Tissue Res. 260: 529-533.

McCormick, SD (1990b) Cortisol directly stimulates differentiation of chloride cells in tilapia opercular membrane. Am. J. Physiol. 259: R857-R863.

McCormick, SD, Hasegawa, S, and Hirano, T (1992) Calcium uptake in the skin of a freshwater teleost. Proc. Natl. Acad. Sci. USA 89: 3635-3638.

Perry, SF, and Flik, G (1988) Characterisation of branchial transepithelial calcium fluxes in freshwater trout, *Salmo gairdneri*. Am. J. Physiol. 254: R491-R498.

Perry, SF, Goss, GG and Fenwick, JC (1992) Interrelationship between gill chloride cell morphology and calcium uptake in freshwater teleosts. Fish Physiol. Biochem. 10: 327-337

Schoenmakers, ThJM, Klaren, PHM, Flik, G, Lock, RAC, Pang, PKT, and Wendelaar Bonga, SE (1992). Actions of cadmium on basolateral plasma membrane proteins involved in calcium uptake by fish intestine. J. Membrane Biol. 127: 161-172.

Verbost, PM, Flik, G, Lock, RAC, and Wendelaar Bonga, SE (1987) Cadmium inhibition of Ca^{2+} uptake in rainbow trout gills. Am. J. Physiol. 253: R216-R221.

Verbost, PM, Van Rooij, J, Flik, G, Lock, RAC, and Wendelaar Bonga, SE (1989) The movement of cadmium through freshwater trout branchial epithelium and its interference with calcium transport. J. Exp. Biol. 145: 185-197.

Verbost, PM, Flik, G, Pang, PKT, Lock, RAC, and Wendelaar Bonga, SE (1989) Cadmium inhibition of the erythrocyte Ca^{2+} pump. J. Biol. Chem. 264: 613-5615.

Verbost, PM, Flik, G, Fenwick, JC, Greco, AM, Pang, PKT, and Wendelaar Bonga, SE (1993) Branchial calcium uptake: possible mechanisms of control by stanniocalcin. Fish Physiol. Biochem. 11:205-125.

Verbost, PM, Schoenmakers, ThJM, Flik, G and Wendelaar Bonga, SE (1994) Kinetics of ATP- and Na^+-gradient driven Ca^{2+} transport in basolateral membranes from gills of freshwater- and sea water-adapted tilapia. J. exp. Biol. 186, 95-108.

Verbost, PM, Bryson, SE, Wendelaar Bonga, SE and Marshall, WS (1996) Na^+-dependent Ca^{2+} uptake in isolated opercular epithelium of *Fundulus heteroclitus*. J. Comp. Physiol. B: in press.

Wendelaar Bonga, SE, L`wik, CJM and van der Meij, JCA (1983) Effects of external Mg^{2+} and Ca^{2+} on branchial osmotic water permeability and prolactin secretion in the teleost fish *Sarotherodon mossambicus*. Gen. Comp. Endocrinol. 52: 222- 231.

Drinking in marine, euryhaline and freshwater teleost fish

J. Fuentes and F.B. Eddy

Department of Biological Sciences, University of Dundee, DD1 4HN, Dundee, Scotland, UK

9.1. The renin angiotensin system: general overview

The renin angiotensin system (RAS) has been extensively studied in mammals and has been the model for comparative studies in fish. All the components of the RAS have been described in teleost fish, although their complete biological actions are not yet fully understood. Recently Olson (1992) produced a comprehensive paper reviewing cardiovascular effects of the RAS in fish in relation to extracellular volume regulation. Therefore, this chapter will focus on the role of the RAS in control of drinking in teleosts in the context of osmoregulation. Topics reviewed herein include drinking in larvae and adults in freshwater, euryhaline and marine fish.

The proteolitic activity of renin initiates a cascade by cleaving the decapeptide angiotensin I (Ang I) from the precursor molecule angiotensinogen. In a further step the angiotensin converting enzyme (ACE) produces the octapeptide angiotensin II (Ang II), by hydrolysis of a dipeptide in the carboxyl terminal end of Ang I. The production of the heptapeptide angiotensin III (Ang III) and the hexapeptide angiotensin IV (Ang IV) are also mediated by peptidases (see Figure 9.1 for further details).

In contrast to the initial concep of a single systemic RAS, in the last few years, some tissue RASs have been described. In mammals, local systems have been identified in brain, kidney, heart, blood vessels and reproductive tissues (Campbell, 1987; Anderson *et al.*, 1995; Unger *et al.*, 1988; Saavedra, 1992; Bottari *et al.*, 1993). Although the whole process is not fully understood, the systemic RAS may affect tissue Ang II levels by providing ACE, angiotensinogen or renin (Johnston, 1990). It seems clear that the production and release of renin plays a key role in the actions of the RAS since it is the limiting/regulating factor of Ang II synthesis (Ondetti and Cushman, 1982).The primary mechanisms regulating the secretion of renin have been identified at the cellular and subcellular level in mammals and have extensively reviewed by King *et al.* (1993).

As far as it is known Ang I is biologically inactive (Olson, 1992). Numerous studies have shown an important role of circulating Ang II in the control of blood pressure, and electrolyte and fluid homeostasis (see reviews by Fitzsimons, 1980; Phillips, 1987 and Andersson *et al.*, 1995). The heptapeptide Ang III and the hexapeptide Ang IV may elicit similar actions and be of biological importance in mammals (Wright and Harding, 1994; Head and Williams, 1992). Apart from the fact that Ang III is not dipsogenic in eels (Hirano and Hasagawa, 1984) nothing is known about the presence or biological actions of Ang III or IV in fish.

In mammals Ang II receptors show a high degree of heterogeneity. The use of peptide (e.g. saralasin) and non-peptide (e.g. losartan) receptor antagonists has revealed the existence of two major subtypes AT_1 and AT_2. Saralasin prevents the binding of Ang II to both subtypes, while losartan specifically antagonises the AT_1 subtype (Smith *et al.* 1992; Brunner *et al.* 1993; Wright and Harding, 1994 and Naveri, 1995).

The presence of Ang II binding has been demonstrated in a series of different tissues in freshwater and seawater adapted trout (Cobb and Brown, 1992) and described in more detail in the glomerulous of the trout (Cobb and Brown, 1993), where angiotensin II has an antidiuretic action (Brown *et al.*, 1980; Gray and Brown, 1985). Furthermore Marsigliante *et al.* (1994), described two different subtypes of receptor in seawater eels, although their physiological actions remain obscure.

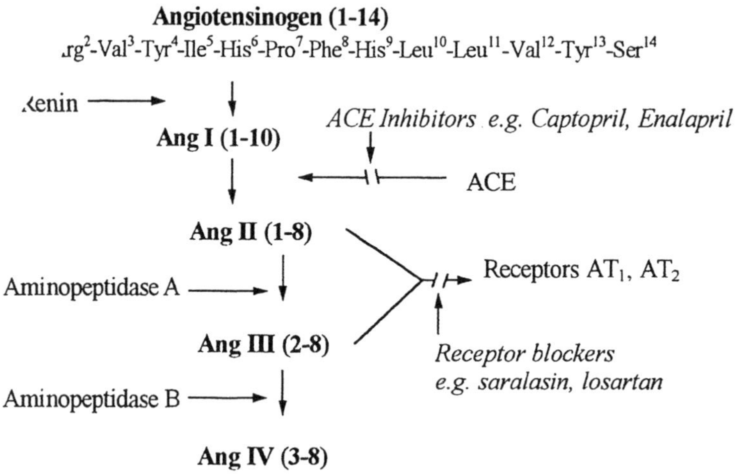

Fig. 9.1. The renin angiotensin system: enzymes, peptides, inhibitors and receptors. Numbers in brackets accompanying each form of angiotensin represent the sequence with respect to angiotensinogen. (See Phillips, 1987; Andersson *et al.*, 1995 and Naveri, 1995 for further details). Details of Ang II sequence are given in Table 9.1.

In mammals, Ang II stimulates thirst and Na^+ appetite (Davies *et al.*, 1970; Fitzsimons, 1980; Nishimura, 1987). In addition to peripheral actions, the intracerebroventricular (i.c.v.) administration of Ang II invokes the pressor response and the stimulation of drinking (Phillips, 1987). All the components of the RAS have been identified in the brain of mammals and the central RAS has been subject of several studies in the last two decades. It seems that the central and peripheral RASs interact in the control of blood pressure and the drinking response, however the actual physiological regulation of thirst by Ang II remains unclear (Phillips, 1987 and Andersson *et al.*, 1995 for extensive reviews). To date the presence of a brain RAS in fish remains an unanswered question, and little is known about the effect of central administration of different components of the RAS. Le Mevel *et al.* (1994) reported, however, hypertensive effects of centrally administered Ang II in i.c.v. cannulated rainbow trout.

Table 9.1. Amino acid substitutions in the sequence of fish angiotensins compared to mammals.

Species	Ang II								Ang I	
	1	2	3	4	5	6	7	8	9	10
[1]Human, rat, rabbit, pig	Asp	Arg	Val	Tyr	Ile	His	Pro	Phe	His	Leu
[2]Goosefish	Asn				Val				His	
[3]Salmon	Asn/Asp				Val				Asn	
[4]Eel	Asp/Asn				Val				Gly	

Substitutions of Asp/Asn in position 1 may be caused by in vitro conversions Khosla *et al.* (1985). References: (1) Akagi *et al.* (1982), (2) Hayashie *et al.* (1978), (3) Takemoto *et al.* (1983) and (4) Hasegawa *et al.* (1983)

9.2. The renin angiotensin system and drinking in teleost fish.

Smith (1930), in one of the most important contributions to the understanding of fish osmoregulation, proposed that seawater adapted fish were able to drink the surrounding water to replace osmotic losses ocurring through the body surfaces. Using phenol red as a non absorbable marker he demonstrated drinking in seawater fish, claiming that ligation of the oesophagus prevented intake of the dye, which otherwise accumulated in the gut to a concentration 3 or 4 times higher that in the medium. It took almost forty years until Evans and Potts (Evans, 1967; Evans and Potts, 1967; Evans, 1968) proposed new methods for a more accurate measurement of drinking in fish using non absorbable radiolabelled compounds.

It has been suggested that the RAS is the main endocrine control of drinking in fishes and numerous reports demonstrated the dipsogenic response by administration of exogenous peptides, or by stimulation of the endogenous RAS (see Table 9.2 for references). However, administration of exogenous human angiotensins failed to stimulate drinking is some stenohaline fish (Kobayashi *et al.* 1983).

During the 80's some fish peptides were purified and sequenced demonstrating some amino acid substitutions in fish sequences with respect to the mammalian peptide (see table 9.1 for details), and administration of homologous peptides results in a more pronounced dipsogenic response in fish.

9.3. Drinking in freshwater and role of the RAS.

It has long been assumed that drinking in freshwater fish is at a very low level or a non-existant process (Smith, 1930; Shehadeh and Gordon, 1969). However, reports are accumulating giving evidence of drinking in freshwater adapted fish (Perrot *et al.*, 1992; Tierney *et al.*, 1995a; Fuentes *et al.*, 1996a). There seems to be an important size dependent effect on drinking rates in freshwater adapted fish (see Figure 9.2) although the physiological meaning remains unknown.

As in mammals, there seems to be an important link between the regulation of blood pressure and drinking in fish. Administration of papaverine, a smooth muscle relaxant, is believed to stimulate drinking by causing hypotension and activation of the endogenous RAS (Balment and Carrick, 1985; Tierney *et al.*,

1995a,b). The stimulation of the RAS results in an increase of circulating levels of Ang II inhibitable by ACE, as demonstrated by Tierney *et al.* (1995a,b) in eels. The relationship between drinking and hypotension in Atlantic salmon alevins has also been shown by administration of sodium nitroprusside either by exposure or injection, since sodium nitroprusside is a nitric oxide releasing agent known to induce hypotension in trout (McGeer and Eddy, 1996). The increased drinking induced by sodium nitroprusside could be abolished by inhibition of ACE by injection of enalapril (Fuentes *et al.*, 1996a)

Table 9.2. The effect of administration of angiotensins on drinking rates in some species of freshwater, euryhaline and seawater fish.

Species	∞C	Salinity	Dose	Angiotensin	Action
Fundulus heteroclitus	15	FW	0.1μg/fish[a]	N.S. A II	drinking[1]
	20	SW	0.1μg/fish[a]	N.S. A II	drinking[1]
Cottus bairdy	14-18	FW	3-10ng/g[c]	Asn^1, Val^5 A II	no action[2]
	14-18	SW	13μg/g[c]	Asn^1, Val^5 A II	drinking[2]
Carassius auratus	25	FW	1.5-4μg/100g[a]	Asp^1, Ile^5 A II	no action[2]
	25	FW	0.1μg/g[a]	Asn^1, Val^5 A II	drinking[3]
Anguilla japonica	18	FW	0.5ng/g[b]	Asp^1, Ile^5 A II	drinking[4]
	18	FW	0.5ng/g[b]	Asp^1, Ile^5 A I	drinking[4]
Anguilla anguilla	10-12	FW	0.3μg/g[c]	Asn^1, Val^5 A II	drinking[5]
	10-12	SW	0.3μg/g[c]	Asn^1, Val^5 A II	drinking[5]
Platichthys flessus	8-10	FW	0.3μg/g[c]	Val^5 A I	drinking[5]
	8-10	SW	6, 150ng/min[d]	Asp^1, Val^5 A II	drinking[6]
Oncorhynchus	12-15	FW	0.1-2μg/g[c]	Asn^1, Val^5 A I	drinking[7]
mykiss	12-15	FW	0.2-2μg/g[c]	Asn^1, Val^5 A II	drinking[7]
Salmo salar	12	FW	0.5μg/g[a]	Asn^1, Val^5 A II	drinking[8]
Sarotherodon mossambicus	25	FW	0.01μg/g[c]	Asn^1, Val^5 A II	no action[3]
Cyprinus carpio	10-12	FW	0.3μg/g[c]	Val^5 A I	drinking[5]
Myxocephalus scorpius	10-12	SW	0.3μg/g[c]	Val^5 A I	no action[5]

(N.S).sequence not specified. (a) intraperitoneal injection. (b) intraarterial bolus injection. (c) intramuscular injection. (d) infusion. References: (1) Malvin *et al.* (1980); (2) Beasley *et al.* (1986); (3) Kobayashi *et al.* (1983); (4) Hirano and Hasegawa (1984); (5) Perrot *et al.* (1992); (6) Carrick and Balment (1983); (7) Fuentes and Eddy (1996) (8) Fuentes and Eddy (unp).

The mechanism by which hypotension enhances drinking is however not fully understood. Hirano and Hasegawa (1984) demonstrated inhibition of drinking during hypertensive episodes (of about 10 minutes) following the administration of exogenous Ang II in freshwater eels and drinking resumed after about 15 minutes, when the hypertensive effect of Ang II disappeared. This result is in keeping with our findings (Fuentes and Eddy, unpublished). A bolus injection of 25μg/kg of [Asn^1, Val^5] Ang II to cannulated freshwater trout resulted in a short hypertensive response (about 15 minutes), followed by a long lasting hypotensive effect (between 30-60 minutes postinjection), an adequate stimulus for initiation of drinking. This response could be abolished by co-administration of saralasin which had been previously demonstrated to inhibit angiotensin-dependent drinking in freshwater trout (Fuentes and Eddy, 1996).

The actual physiological role of drinking in freshwater fish remains unexplained, and it has been suggested that in alevins and larvae, drinking might be involved in calcium uptake or some aspect of feeding (Tytler *et al.*, 1990).

The involvement of the RAS in control of drinking is clear and as shown above the stimulation of the endogenous RAS or the administration of RAS components to freshwater adapted fish (Fuentes and Eddy, 1996), results in enhanced drinking. However, ACE inhibitors administered to freshwater adapted fish are unable to decrease basal drinking (Tierney *et al.*, 1995a; Fuentes and Eddy, 1996), suggesting the possibility of an additional hormonal system with a role in control of drinking in freshwater fish.

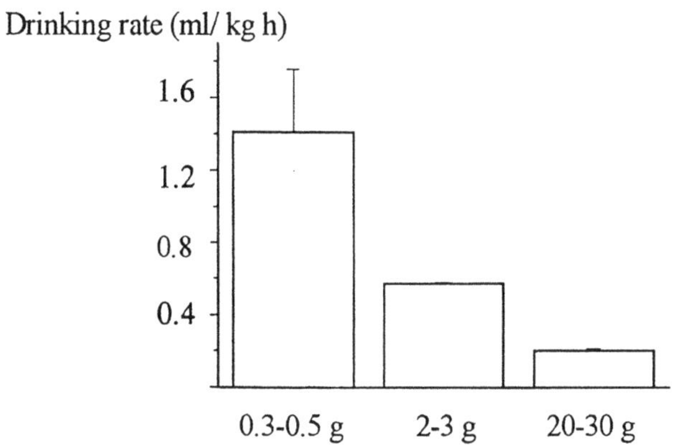

Fig. 9.2. Drinking rates in juvenile Atlantic salmon adapted to freshwater: Effect of fish size. Each column represents the mean±SEM of 12-15 fish.

9.4. Drinking in seawater

Regulation of drinking is essential in teleosts moving from freshwater to seawater or *vice versa,* and euryhaline fish face different osmoregulatory problems depending on the salinity of their environment. In freshwater, the osmoregulatory processes are focused on ion uptake and water removal from the body. Upon entry in seawater the removal of the excess salts and the replacement of water lost due to dehydration of the body surfaces (mainly the gills) are essential to maintain the homeostasis of body fluids.

The transfer of fish from freshwater to higher salinities results in an increase of drinking (see table 9.3). For instance, drinking rates in the flounder are higher in seawater fish than in fish kept in fresh water (Balment and Carrick, 1985), and the transfer of the European eel (*Anguilla anguilla*) from freshwater to seawater results in a 10 fold increase of drinking rates (Tierney *et al.*, 1995a).

As in freshwater adapted fish, drinking in seawater fish is dependent upon an intact RAS and the inhibition of ACE with specific blockers (i.e. captopril) results in the decrease of the basal drinking rate in sea water (Tierney *et al.*, 1995a). Plasma renin activity is higher in seawater than in freshwater adapted eels (Sokabe *et al.*, 1973; Henderson *et al.*, 1976), and Ang II levels in seawater eels are also higher than in freshwater adapted fish (Tierney *et al.*, 1995a).

Table 9.3. Drinking rates in some representative marine and euryhaline fish adapted to seawater.

Specie	BW (g)	Drinking rate	Reference
Xiphister atropurpureus	-	8.14 ml/kg day	Evans (1968)
Pholis gunnelus	2-10	12.33 ml/kg day	Evans (1967)
Serranus scriba	150	563 μl/h 100g [a] 461 μl/h 100g [b] 259 μl/h 100g [c]	Isaia (1972)
Aphanius dispar	0.4-1	11 μl/g h[a] 5.6 μl/g h[d]	Skadhauge and Lotan (1974)
Fundulus heteroclitus	-	0.6 % body wt/h	Malvin *et al.* (1980)
Trachurus japonicus	59-120	13.6 μl/h 10g	Kobayashi *et al.* (1983)
Mugil cephalus	18-30	82.6 μl/h 10g	Kobayashi *et al.* (1983)
Cottus bairdi	0.5-7.8	0.09 % body wt/h	Beasley *et al.* (1986)
Platichtys flessus	150-400	240 μl/h 100g	Balment and Carrick (1985)
Salmo salar (smolts)	20-30	7 ml/kg h	Usher *et al.* (1988)
Limanda limanda	-	3.6 ml/kg h	Perrot *et al.* (1992)
Myxocephalus scorpius	-	7.76 ml/kg h	Perrot *et al.* (1992)
Scophthalmus maximus	45-200	0.08 ml/100g h	Carrol *et al.* (1994)
Anguilla anguilla	250-800	0.51 ml/kg h	Tierney *et al.*(1995a)
Oncorhynchus mykiss	20-30	2.4 ml/kg h	Fuentes *et al* (1996b)

(BW) body weight, (a) 20 ∞C, (b) 15 ∞C (c) 10 ∞C (d) 17 ∞C

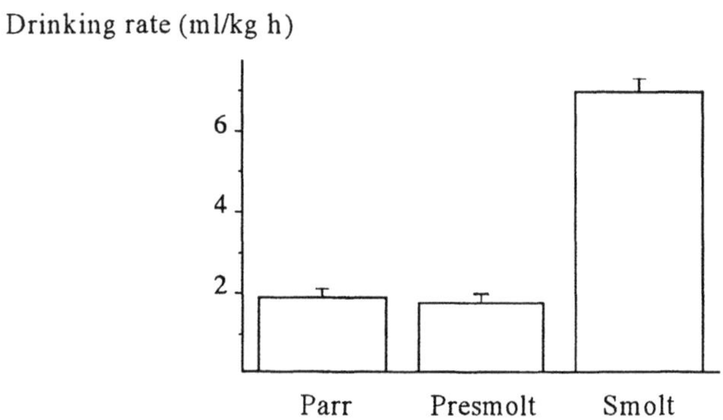

Fig. 9.3. Drinking rate in juvenile Atlantic salmon 7 days after transfer to seawater at different stages of smoltification. Parr (Smith *et al.*, 1991); presmolts (Fuentes *et al.*, 1996) and smolts (Usher *et al.*, 1988).

The transfer of juvenile Atlantic salmon (*Salmo salar*) to seawater results also in significant increase of drinking. In this case however, a seasonal change is well documented: Atlantic salmon smolts drink to a rate of about 7 ml/kg h after a week in seawater (Usher *et al.*, 1988), while salmon parr (Smith *et al.*, 1991) or presmolts (Fuentes *et al.*, 1996b) develop drinking to a maximum of only 2 ml/kg h (Figure 9.3).

The production of renin in mammals is the limiting factor for Ang II release (Ondetti and Cushman, 1982). In Atlantic salmon, plasma renin activity (PRA) is higher in smolts than in parr after 4 days in seawater, although when both are kept in freshwater before transfer, PRA remains at the same level (Smith *et al.*, 1991). Whether this fact has a direct relationship with salmon smoltification, remains unknown. However, some mammalian studies have revealed seasonal patterns in PRA and in the responsiveness of intestinal preparations to exogenous Ang II (Reinberg *et al.*, 1978; Levens, 1985).

9.5. Drinking in Larvae

Although, protected by the chorion, the main barrier separating marine eggs from the environment is the vitelline membrane (Loeffler and Lovtruo, 1970), and under this protection eggs have a very low permeability to water (Potts and Eddy, 1973). However, after hatching, the vitelline membrane is no longer present and the larvae have to osmoregulate to cope with the external salinity.

Larvae from marine fish studied so far are capable of active drinking as shown in cod, plaice and herring (Mangor-Jensen and Adolf, 1987; Tytler and Blaxter 1988a), halibut (Tytler and Blaxter, 1988b), herring and turbot (Tytler and Ireland, 1994).

Tytler and Blaxter (1988) demonstrated that cod larvae were able to modify drinking rates in response to changes in the external salinity. Furthermore, the accumulation of the non absorbable ^{3}H-dextran in the gut also showed that herring larvae a few days old were able to absorb water as a part of their osmoregulatory process (Tytler and Blaxter, 1988). In addition, Tytler and Bell (1989) showed that cod larvae were able to keep their body fluids hypotonic to seawater in a similar way to adults. Drinking seems to play a key role in osmoregulation in marine larvae, since they are able to drink before the development of a functional anus as it is the case of halibut (Tytler and Blaxter, 1988b), or lack of functional gills, as it is the case of herring (Tytler and Blaxter, 1988a).

Freshwater larvae deserve particular mention, since newly hatched rainbow trout larvae imbibe water to a rate of about 3.2 µl/g h (Tytler *et al.*, 1990), 10 fold higher that adult fish (see Figure 9.2 for comparison). It has been suggested that such high rate of drinking in yolk-sac larvae in freshwater, could be involved in calcium uptake in a stage that gill chloride cells are unable of net calcium uptake (Tytler *et al.*, 1990). The actual physiological meaning of drinking in freshwater larvae and adults remains unknown and, further research is needed to address this interesting issue.

9.6. Absorption of water and salts in marine fish

The oesophagus plays a key role in the processing of ingested water. While it is highly impermeable to water and salts in freshwater fish, in seawater fish the imperviousness to water remains but it is the major site for net salt uptake. The oesophagus produces a true desalination of ingested seawater without loss of water from serosa to lumen. Thus the imbibed fluid moves along the intestine and can be absorbed following the absorption of salts driven by a Na-K-Cl cotransport (Munch *et al.*, 1982).

Kirsch and Maister (1982) postulated that water influx in the eel gut takes place as a function of transepithelial chloride concentration and using a different approach, Ando and Nagashima (1996) suggested that high concentrations of Cl^- in the gut occurring as a result of poor desalination in the oesophagus, were able to inhibit water intake, that was resumed only when chloride concentration descended below a certain level.

The effect of Ang II in the process of absorption/secretion of fluid in the mammal intestine is clear (Levens, 1985). Though the involvement of Ang II in water absorption in the gut remains unstudied in fish, indirect evidence has been obtained by the stimulation of the endogenous renin angiotensin system which increases drinking and results in the decrease of plasma osmolarity in freshwater fish (Balment and Carrick 1985; Tierney *et al.*, 1995a), supporting the idea of the involvement of Ang II in water absorption in the gut.

Atrial natriuretic peptide has been shown to inhibit the intestinal Na-K-Cl cotransport in seawater adapted flounder (O'Grady *et al.*, 1985), intestinal transport in preparations of goby intestine (Loretz, 1995) and Ando *et al.* (1992) reported a decrease in NaCl and water absorption *in vitro* in response to heterologous (human) or homologous (eel) ANPs. However, as suggested by Loretz (1995) EC_{50} values for ANP of 7 nM (O'Grady *et al.*, 1985), 10 nM (Ando *et al.*, 1992) and 156 nM (Loretz, 1995) are about 200-fold greater than plasma ANP content in seawater adapted fish and compromise the physiological effect of the peptide *in vivo*.

9.7. Effect of cortisol on drinking.

Cortisol and growth hormone have a well known potential to improve the adaptation of salmonids to seawater, reducing the rate of mortality and enhancing hypoosmoregulation (Madsen, 1990a; Sakamoto *et al.*, 1993). Stimulation of gill and intestinal Na^+-K^+-ATPase are the most relevant effects of cortisol in terms of osmoregulation (Madsen, 1990b; Boeuf *et al.*, 1994). Water absorption in gut preparations of Atlantic salmon smolts adapted to fresh water or sea water was unaffected by cortisol (Usher *et al.*, 1991). The administration of cortisol to freshwater adapted rainbow trout and Atlantic salmon presmolts (Fuentes *et al.*, 1996) improves the hypoosmoregulatory performance and increases drinking rates after the transfer to seawater. Atlantic salmon presmolts pre-treated with ovine growth hormone (Figure 9.4) show the same response as noted with cortisol following seawater transfer (Fuentes and Eddy, unpublished). It is remarkable however, that neither cortisol nor growth hormone induced an increase in drinking rate while fish were in freshwater.

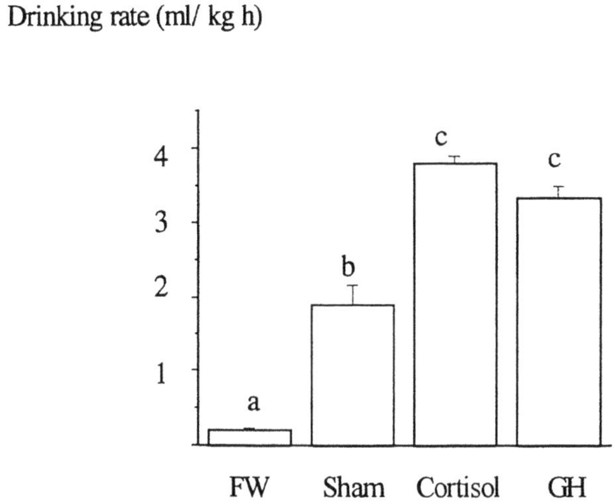

Fig. 9.4. Drinking rate in Atlantic salmon presmolts (30-40 g) in freshwater and after 7 days in seawater. Each column represents the mean± SEM of 6-7 fish. Significant differences ($p<0.05$, one-way Anova) are denoted by different letters. Cortisol group received an implant of 80mg/kg in coconut oil and were kept in freshwater for two weeks before transfer to seawater. Growth hormone (GH) treated group received a daily injection of hormone for a week before transfer (50μg oGH/fish). Sham group received either saline injections or implants of coconut oil alone. Cortisol results from Fuentes *et al.* (1996).

9.8. Effect of Atrial Natriuretic Peptide on drinking.

The existence of bioactive atrial peptides with effects on fluid and electrolyte homeostasis was first suggested by de Bold *et al.* (1981). Atrial extracts from rats were able to induce an important natriuresis and diuresis in test animals. Since then, natriuretic peptides have been purified and sequenced in some mammalian species (Brenner *et al.*, 1990) and fish (Takei *et al.*, 1989).

ANP has been shown to have vascular and cardiac effects and some renal and extrarenal osmoregulatory actions in fish and Evans (1990), Olson (1992) and Takei (1993) have produced timely reviews of most of the known actions of natriuretic peptides in fish.

In mammals, the adaptive significance of ion transport regulation by ANP is well known, and ANP induces an inhibition of net salt/fluid absorption (Loretz, 1995), in response to hypertension, hypervolemia or excess salt intake (Cantin and Genest, 1985; Gutkowska and Nemer, 1989). In fish, intestinal salt and water uptake are coupled (Musch *et al.*, 1982) and the inhibitory effect of ANP in the absorptive process (O'Grady *et al.*, 1985) would compromise the hydration of the fish, mainly in seawater, and result in maladaptative responses (see Evans, 1990). Atrial natriuretic peptide has inhibitory effects on drinking in eels *in vivo* (Takei and Balment, 1993; Tierney *et al.*, 1996) and water and NaCl absorption *in vitro* (Ando *et al.*, 1992).

9.9. Effect of temperature on drinking

The effect of temperature on osmoregulation has been extensively studied in fish. In freshwater the adaptation to very low temperatures usually induces a decrease of blood osmolarity, while marine fish subjected to the same conditions show the opposite response.

Drinking in fish adapted to a variety of temperatures within their normal living range are usually positively related with temperature showing Q_{10} values of about 2. Seawater adapted flounders increase their drinking responses when adapted to higher temperatures (Maetz and Evans, 1972). Positive relationships between temperature and drinking have also been reported in eels adapted to seawater (Motais and Isaia, 1972) or the euryhaline *Aphanius dispar* adapted to seawater (Skadhauge and Lotan, 1974). The effect of temperature on drinking remains unstudied in freshwater adapted fish.

Acute and chronic adaptations to temperature changes may have different species specific effects. The acute transfer of flounders from 7 to 20 ∞C results in increase of drinking and induces an extensive mortality in turbot. However, the transfer of turbot from 13 to 20 ∞C, improves the survival but results in a decrease of drinking rates (Carroll *et al.*, 1994). Chronic changes in temperature (i.e. seasonal increase from winter to summer) has also an effect on drinking, and two species of flatfish (i.e. dab and plaice) studied by Carroll *et al.* (1995) showed higher drinking rates during the summer.

References

Akagi, H., Hayashi, T., Nakayama, T., Nakayima, T., Watanabe, T.X. and Sokabe, H. 1982. Comparative studies in angiotensins. VI Structure of angiotensin I produced by renal renin of the dog, guinea pig, and rabbit, and re-examination of the peptides of the pig, horse, and ox using homologues renin sources. Chem. Pharm. Bull. 30:2498-2510.

Andersson, B., Eriksson, S. and Rundgren, M. 1995. Angiotensin and the brain. Acta Physiol. Scand. 155: 117-125.

Ando, M., Kondo, K. and Takei, Y. 1992. Effects of eel atrial natriuretic peptide on NaCl and water transport across the intestine of the seawater eel. J. Comp. Physiol. B 162: 436-439.

Ando, M. and Nagashima, K. 1996. Intestinal Na^+ and Cl- levels control drinking behaviour in the seawater-adapted eel *Anguilla japonica*. J. exp. Biol. 199: 711-716.

Balment, R.J. and Carrick, S. 1985. Endogenous renin-angiotensin system and drinking behaviour in flounder. Am. J. Physiol. 248: R157-160.

Balment, R.J., Takei, Y., Tierney, M.L. and Hazon, N. 1996. Atrial natriuretic peptide and angiotensin involvement in the control of drinking in sea water-adapted eels, *Anguilla anguilla*. J. Endocrinol. 147: (Suppl) P90.

Beasley, D., Shier, D.N., Malvin, R.L. and Smith, G. 1986. Angiotensin-stimulated drinking in marine fish. Am. J. Physiol. 250: R1034-R1038.

Boeuf, G., Marc, P., Prunet, P., LeBail, P.Y. and Smal, J. 1994. Stimulations of parr-smolt transformations by hormonal treatment in Atlantic salmon (*Salmo salar* L.). Aquaculture 112: 63-68.

Bottari, S.P., de Gasparo, M., Steckelings, U.M. and Levens, N.R. 1993. Angiotensin II receptor subtypes: characterisation, signalling mechanisms, and possible physiological implications. Front Neuroendocrinol. 14: 123-171.

Brenner, B.M., Ballermann, B.J., Gunning, M.E. and Zeidel, M.L. 1990. Diverse biological actions of atrial natriuretic peptide. Pharmacol. Rev. 70: 665-669

Brown, J.A., Oliver, J.A., Henderson, I.W. and Jackson, B.A. 1980. Angiotensin and single nephron function in the trout *Salmo gairdneri.* Am. J. Physiol. 239: R509-R514.

Brunner, H.R., Nussberger, J. and Waever, B. 1993. Angiotensin antagonists. In: Robertson, J.I.S. and Nichols M.G., (eds.) *The renin-angiotensin system: Biochemistry, Physiology, Pathophysiology, Therapeutics:* Vol. 1. London: Gower Medical Publisher: 35.1-35.14

Campbell, D.J. 1987. Circulating and tissue angiotensin systems. J. Clin. Invest. 79: 1-6.

Cantin, M. and Genest, J. 1985. The heart and the atrial natriuretic peptide. Endocr. Rev. 6: 107-127.

Carrick, S. and Balment, R.J. 1983. The renin-angiotensin system and drinking in the euryhaline flounder, *Platichthys flessus.* Gen. Comp. Endocrinol. 51: 423-433.

Carroll, S., Kelsall, C., Hazon, N. and Eddy, F.B. 1994. Effect of temperature on the drinking rates of two species of flatfish, flounder and turbot. J. Fish Biol. 44: 1097-1099.

Carroll, S., Hazon, N. and Eddy, F.B. 1995. Drinking rates and Na^+ effluxes in response to temperature change in two species of marine flatfish: dab, *Limanda limanda* and plaice, *Pleuronectes platessa.* J. Comp. Physiol. B, 164: 579-584.

Cobb, C.S. and Brown, J.A. 1992. Angiotensin II binding to tissues of the rainbow trout, *Oncorhynchus mykiss,* studied by autoradiography. J. Comp. Physiol. B 162: 197-202.

Cobb, C.S. and Brown, J.A. 1993. Characterisation of putative glomerular receptors for angiotensin II in the rainbow trout *Oncorhynchus mykiss* using the antagonists losartan, PD123177, and saralasin. Gen. Comp. Endocrinol. 92: 123-131.

Davies, N.T., Munday, K.A. and Parsons, B.J. 1970. The effect of angiotensin in rat fluid transport. J. Endocrinol. 48: 39-46.

de Bold, A.J., Borenstein, H.B., Veress, A.T. and Sonnemberg, H. 1981. A rapid and potent natriuretic response to intravenous injection of atrial myocardial extract in rats. Life Sci. 28: 89-94.

Dzau, V.J. 1993. Vascular renin-angiotensin system and vascular protection. J. Cardiovasc. Pharmacol. 22 (suppl 5): S1-S9.

Evans, D.H. 1967. Sodium, chloride and water balance of the intertidal teleost, *Xiphister atropurpureous-* II. The role of the kidney and the gut. J. exp. Biol. 47: 519-524.

Evans, D.H. 1968. Measurement of drinking rates in fish. Comp. Biochem. Physiol. 25: 751-753.

Evans, D.H. 1990. An emerging role for a cardiac peptide hormone in fish osmoregulation. Ann. Rev. Physiol. 52: 43-60.

Fitzsimons, J.T. 1980. Angiotensin and other peptides in the control of water and sodium intake. Proc. R. Soc. Lond. 210:165-182.

Fuentes, J. and Eddy, F.B. 1996. Drinking in freshwater adapted rainbow trout fry *Oncorhynchus mykiss* (Walbaum) in response to angiotensin I, angiotensin II, angiotensin converting enzyme inhibition and receptor blockade. Physiol. Zool. *In press.*

Fuentes, J., McGeer, J.C. and Eddy, F.B. 1996a. Drinking rate in juvenile Atlantic salmon, *Salmo salar* fry in response to a nitric oxide donor, sodium nitroprusside and an inhibitor of angiotensin converting enzyme, enalapril. Fish Physiol. Biochem. 15: 65-69.

Fuentes, J., Bury, N.R., Carrol, S. and Eddy, F.B. 1996b. Drinking in Atlantic salmon presmolts (*Salmo salar* L.) and juvenile rainbow trout (*Oncorhynchus mykiss* Walbaum) in response to cortisol and sea water challenge. Aquaculture. 141: 129-132.

Gray, C.J. and Brown, J.A. 1985. Renal and cardiovascular effects of angiotensin II in the rainbow trout, *Salmo gairdneri.* Gen. Comp. Endocrinol. 59: 375-381.

Gutkowska, J. and Nemer, M. 1989. Structure, expression, and function of atrial natriuretic factor in extraatrial tissues. Endocr. Rev. 10: 519-536.

Hasegawa, Y., Nakayima, T. and Sokabe, H. 1983. Chemical structure of angiotensin formed with kidney renin in the Japanese eel, *Anguilla japonica.* Biomed. Res. 4: 417-420.

Hayashi, T., Nakayama, T., Nakayima, T. and Sokabe, H. 1978. Comparative studies on angiotensins. V. Structure of angiotensin formed by the kidney of Japanese goosefish and its identification by Dansyl method. Chem. Pharm. Bull. 26: 215-219.

Head, G.A. and Williams, N.S. 1992. Hemodynamic effects of central angiotensin I, II and III in conscious rabbits. Am. J. Physiol. 263: R845-R851.

Henderson, I.W., Jotisankasa, V., Hosley, W. and Oguri, M. 1976. Endocrine and environmental influences upon plasma cortisol concentrations and plasma renin activity of the eel, *Anguilla anguilla.* J. Endocrinol. 70: 81-95.

Hirano, T. and Hasegawa, S. 1984. Effects of angiotensins and other vasoactive substances on drinking in the eel, *Anguilla japonica.* Zool. Sci. 1: 106-113.

Isaia, J. 1972. Comparative effect of temperature on the sodium and water permeabilities of the gills of a stenohaline freshwater fish (*Carassius auratus*) and the stenohaline marine fish (*Serranus scriba, Serranus cabribus*). J. exp. Biol. 57: 359-366.

Johnston, C.I. 1990. Biochemistry and pharmacology of the renin-angiotensin system. Drugs 39 (suppl 1): 21-31.

Khosla, M.C., Nishimura, H., Hasegawa, Y. and Bumpus, F.M. 1985. Identification and synthesis of (1-Asparagine, 5-valine, 9-glycine) angiotensin I produced from the plasma of American eel *Anguilla rostrata.* Gen. Comp. Endocrinol. 57: 223-233.

King, J.A., Lush, D.J. and Fray, J.C.S. 1993. Regulation of renin processing and secretion: chemiosmotic control and novel secretory pathway. Am. J. Physiol. 265: C305-C320.

Kirsch, R. and Meister, M.F. 1982. Progressive processing of ingested water in the gut of sea-water teleosts. J. exp. Biol. 98: 67-81.

Kobayashi, H., Uemura, H., Takei, Y., Itatsu, N., Ozawa, M. and Ichinohe, K. 1983. Drinking induced by angiotensin II in fishes. Gen. Comp. Endocrinol. 49: 295-306.

Le Mevel, J.C., Pamantung, T.F., Mabin, D. and Vaudry, H. 1994. Intracerebroventricular administration of angiotensin II increases heart rate in the conscious trout. Brain Res. 654: 216-222.

Levens, N.R. 1985. Control of intestinal absorption by renin-angiotensin system. Am. J. Physiol. 249: G3-G15.

Loeffler, C.A. and Lovtrup, S.L. 1970. Water balance in the salmon egg. J. exp. Biol. 52: 291-298.

Loretz, C.A. 1995. Atrial natriuretic peptide regulation of vertebrate intestinal ion transport. Amer. Zool. 35: 490-502.

Madsen, S.S. 1990a. The role of cortisol and growth hormone in seawater adaptation and development of hypoosmoregulatory mechanisms in sea trout parr *(Salmo trutta trutta)*. Gen. Comp. Endocrinol. 79: 1-11.

Madsen, S.S. 1990b. Cortisol treatment improves the development of hypoosmoregulatory mechanisms in the euryhaline rainbow trout, *Salmo gairdneri*. Fish Physiol. Biochem. 8: 45-52.

Maetz, J. and Evans, D.H. 1972. Effects of temperature on branchial sodium-exchange and extrusion mechanisms in the seawater adapted flounder *Platichthys flessus* L. J. exp. Biol. 56: 565-585

Malvin, R.L., Schiff, D. and Eiger, S. 1980. Angiotensin and drinking rates in the euryhaline killifish. Am. J. Physiol. 239: R31-R34.

Mangor-Jensen, A. and Adoff, G.R. 1987. Drinking activity of newly hatched larvae of cod (*Gadus morhua*). Fish Physiol. Biochem. 3: 99-103.

Marsigliante, S., Verri, T., Barker, S., Jimenez, E., Vinson, G.P. and Storelli, C. 1994. Angiotensin II receptor subtypes in eel (*Anguilla anguilla*). J. Mol. Endocrinol. 12: 61-69.

McGeer, J.C. and Eddy, F.B. 1996. Effects of sodium nitroprusside on blood circulation, acid-base balance and ionic balance in rainbow trout: indications for nitric oxide induced vasodilation. Can J. Zool. *In press*.

Motais, R. and Isaia, J. 1972. Temperature-dependence of permeability to water and sodium of the gill epithelium of the eel, *Anguilla anguilla*. J. exp. Biol. 56: 587-600.

Musch, M.W., Orellana, S.A., Kimberg, L.S., Fiel, M., Halm, D.R., Kransky, E.J. and Fritzzell, R.A. 1982. Na-K-Cl cotransport in the intestine of a marine teleost. Nature 300: 351-354.

Naveri, L. 1995. The role of angiotensin receptor subtypes in cerebrovascular regulation in the rat. Acta Physiol. Scand. 155: sup. 630.

Nishimura, H. 1987. Role of the renin-angiotensin system in osmoregulation. In Pang. P.K.T. and. Schreibman, M.P., (eds.) *Vertebrate Endocrinology: Fundamentals and Biochemical implications*. Vol. 2. San Diego: Academic Press: 157-187

O'Grady, S.M., Field, M., Nash, N.T. and Rao, M.C. 1985. Atrial natriuretic factor inhibits Na-K-Cl cotransport in teleost intestine. Am. J. Physiol. 249: C531-C534.

Olson, K.R. (1992). Blood and extracellular fluid volume regulation: Role of the renin-angiotensin system, kalikrein-kinin system and atrial natriuretic peptides. In: Hoar, W.S., Randall, D.J and. Farrel, A.P. (eds.). *Fish Physiology* Vol XII B. San Diego: Academic Press: 135-254.

Ondetti, M.A. and Cushman, D.W. 1982. Enzymes of the renin-angiotensin system and their inhibitors. Ann. Rev. Biochem. 51: 283-308.

Perrot, M.N., Grierson, C.E., Hazon, N. and Balment, R.J. 1992. Drinking behaviour in sea water and fresh water teleosts, the role of the renin-angiotensin system. Fish Physiol. Biochem. 10: 161-168.

Phillips, M.I. 1987. Functions of angiotensin in the central nervous system. Ann. Rev. Physiol. 49: 4134-435.

Potts, W.T.W. and Evans, D.H. 1967. Sodium and chloride balance in the killifish, *Fundulus heteroclitus*. Biol. Bull. 133: 411-425.

Potts, W.T.W. and Eddy, F.B. 1973. The permeability to water in the eggs of certain marine teleosts. J. Comp. Physiol. 87: 305-315.

Reinberg, A., Lagoguey, M., Casselin, F., Touitou, Y., Legrand, J.C., Delassalle, A., Antreassian, J. and Lagoguey, A. 1978. Circadian and circannual rhythms in plasma hormones and other variables of five healthy young human males. Acta Endocrinol. 88: 417-427.

Saavedra, J.M. 1992. Brain and pituitary angiotensin. Endocr. Rev. 13: 329-380.

Sakamoto, T., McCormick, S.D. and Hirano, T. 1993. Osmoregulatory actions of growth hormone and its mode of action in Salmonids: a review. Fish Physiol. Biochem. 11: 155-164.

Shehadeh, Z.H. and Gordon, M.S. 1969. The role of the intestine in salinity adaptation in the rainbow trout, *Salmo gairdneri*. Comp. Biochem. Physiol. 30: 397-418.

Skadhauge, E. and Lotan, R. 1974. Drinking rate and oxygen consumption in the euryhaline teleost *Aphanius dispar* in waters of high salinity. J. exp. Biol. 60: 547-556.

Smith, H.S. 1930. The absorption and excretion of water and salts by marine teleosts. Am. J. Physiol. 93: 480-505.

Smith, N.F., Eddy, F.B., Struthers, A.D. and Talbot, C. 1991. Renin, atrial natriuretic peptide and blood plasma ions in parr and smolts of Atlantic salmon *Salmo salar* L. and rainbow trout *Oncorhynchus mykiss* (Walbaum) in fresh water and after short-term exposure to sea water. J. exp. Biol. 157: 63-74.

Smith, R.D., Chiu, A.T., Wong, P.C., Herblin, W.F. and Timmermans, P.B.M.W.M. 1992. Pharmacology of nonpeptide angiotensin II receptor antagonists. Ann. Rev. Pharmacol. Toxicol. 32: 135-165.

Sokabe, H., Oide, H., Ogawa, M. and Utida, S. 1973. Plasma renin activity in Japanese eel, *Anguilla japonica*. Gen. Comp. Endocrinol. 21: 160-167.

Takei, Y., Takahashi, A., Watanabe, T.X., Nakajima, K., and Sakakibara, S. 1989. Amino acid sequence and relative biological activity of a natriuretic peptide isolated from eel brain. Biochem. Biophys. Res. Commun. 164: 537-543

Takei, Y. 1993. Role of peptide hormones in fish osmoregulation. In: Rankin, J.C. and Jensen, F.B. (eds.) *Fish Ecophysiology:* London: Chapman and Hall: 136-160.
Takei, Y. and Balment, R.J. 1993. Biochemistry and physiology of a family of eel natriuretic peptides. Fish Physiol. Biochem. 11: 183-188.
Takemoto, Y., Nakajima, T., Hasegawa, Y., Watanabe, T.X., Sokabe, H., Kumabae, S.-I. and Sakakibara, S. 1983. Chemical structures of angiotensins formed by incubating plasma with the kidney and the corpuscles of Stannius in the Chum salmon *Oncorhynchus keta.* Gen. Comp. Endocrinol. 51: 219-227.
Tierney, M.L., Luke, G., Cramb, G. and Hazon, N. 1995a. The role of the renin-angiotensin system in the control of blood pressure and drinking in the European eel, *Anguilla anguilla.* Gen. Comp. Endocrinol. 100: 39-48.
Tierney, M.L., Cramb, G. and Hazon, N. 1995b. Stimulation of the renin-angiotensin system and drinking by papaverine in the seawater eel. J. Fish Biol. 46: 721-724.
Tytler, P. and Bell, M.V. 1989. A study of diffusional permeability of water, sodium and chloride in yolk-sac larvae of cod (*Gadus morhua* L.). J. exp. Biol. 147: 125-132.
Tytler, P . and Blaxter, J.H.S. 1988a. The effects of external salinity on the drinking rates of the larvae of herring, plaice and cod. J. exp. Biol. 138: 1-15.
Tytler, P . and Blaxter, J.H.S. 1988b. Drinking in yolk-sac stage larvae of the halibut, *Hipoglossus hipoglossus* (L.). J. Fish Biol. 32 493-494.
Tytler, P. and Ireland, J. 1994. Drinking and water absorption by the larvae of herring (*Clupea harengus*) and turbot (*Scophthalmus maximus*). J. Fish Biol. 44: 103-116.
Tytler, P., Tatner, M. and Findlay, C. 1990. The ontogeny of drinking in the rainbow trout *Oncorhynchus mykiss* (Walbaum). J. Fish Biol. 36: 867-875.
Unger, T., Badoer, E., Ganten, D., Lang, R.E. and Retting, R. 1988. Brain angiotensin: pathways and pharmacology. Circulation 77 (suppl I): I-40-I-54.
Usher, M.L., Talbot, C. and Eddy, F.B. 1988. Drinking in Atlantic salmon smolts transferred to seawater and the relationship between drinking and feeding. Aquaculture 73: 237-246.
Usher, M.L., Talbot, C. and Eddy, F.B. 1991. Intestinal water transport in juvenile Atlantic salmon (*Salmo salar* L.) during smolting and following transfer to seawater. Comp. Biochem. Physiol. 100A: 813-818.
Wright, J.W. and Harding, J.W. 1994. Brain angiotensin receptor subtypes in the control of physiological and behavioural responses. Neurosci. Behav. Rev. 18: 21-53.

Teleost renal function:regulation by arginine vasotocin and by angiotensins

J.A. Brown[1] and R.J. Balment[2]

[1] Department of Biological Sciences, University of Exeter, Exeter, UK.
[2] School of Biological Sciences, University of Manchester, Manchester, UK

10.1 Introduction

Euryhaline teleosts are able to live in or adapt to a wide range of environmental salinities and maintain relatively constant plasma osmolality. Adaptive changes in renal function play an important part in this euryhalinity (along with changes in gill and gut transport). In dilute media, high rates of urine output eliminate the osmotic influx of water at the gills, whereas in media more concentrated than plasma, urine flow rates are generally reduced, reflecting reduced glomerular filtration rates (Rankin, Henderson & Brown, 1983). Physiological control of these changes in renal function is poorly understood but is likely to involve endocrine and paracrine factors/systems. The recently described atrial natriuretic peptides (Takei, 1994) have been reported to be diuretic and natriuretic in rainbow trout and toadfish (Duff & Olson, 1986; Lee & Malvin, 1987). However, the physiological significance of these observations awaits clarification as these studies employed heterologous peptides and it is unclear how the doses employed relate to the recent first description of circulating levels of these peptides in the Japanese eel (Kaiya & Takei, 1996). The pituitary hormones, arginine vasotocin (AVT) or isotocin and the renin-angiotensin system (acting through the active angiotensin peptides) have long been implicated in the adaptive regulation of teleost kidney function, though this lacks clear definition. Investigation of the circumstances in which release of these hormones is initiated or in which local systems are activated is of importance in understanding their involvement in control of physiological function. A multiplicity of approaches has been applied in recent years from which we are beginning to gain insights into the role that AVT and angiotensins play in the control of renal function in fish.

10.2 Renin angiotensin system (RAS)

The major biochemical steps involved in the complex renin-angiotensin system are shown in Fig 10.1. In the classical systemic RAS, renin, released by the kidney, acts on a substrate, angiotensinogen, produced by the liver, to form angiotensin I, a peptide with little apparent biological action. Angiotensin converting enzyme (ACE) cleaves two amino acids from the carboxy-terminus of angiotensin I to form angiotensin II in the circulation which is physiologically active (Henderson, Brown & Balment, 1993). Although there is some variation in the angiotensin amino acid sequence between the various vertebrate groups (see

Table 10.1) those described in teleost fish thus far are highly conserved. Characteristically the fish peptides have Asn or Asp in position 1 and valine in

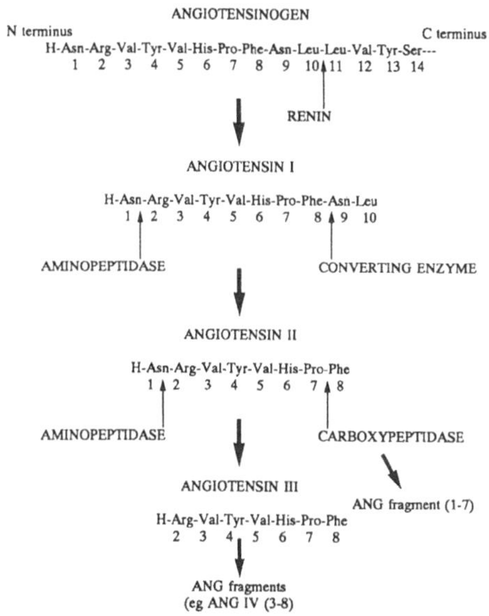

Fig. 10.1. Biochemical steps in the renin-angiotensin system.

position 5, though recently we have isolated flounder angiotensin and discovered that it has Ile in position 5 (Balment & Takei unpublished). There may, therefore, be greater variation in angiotensin structures within the teleosts than previously thought, which has implications for the interpretation of experimental data from fish species where the native sequence has not been determined.

Table 10.1. Amino acid sequences of vertebrate angiotensins

Amino acid number										
N terminus						C terminus				
1	2	3	4	5	6	7	8	9	10	Species
Asp	Arg	Val	Tyr	Val	His	Pro	Phe	His	Leu	Ox,Turtle
Asp	-	-	-	-	-	-	-	His	-	Man, horse, rat rabbit
Asp	-	-	-	Ile	-	-	-	Ser	-	Fowl
Asp/n	-	-	-	Val	-	-	-	Tyr	-	Snake
Asp	-	-	-	Val	-	-	-	Asn	-	Bullfrog
Asn	-	-	-	Val	-	-	-	Asn	-	Salmon
Asn	-	-	-	Val	-	-	-	His	-	Japanese goosefish
Asn	-	-	-	Ile	-	-	-	Thr	-	Flounder

Intravenous infusion of angiotensin II in the anesthetized rainbow trout has an antidiuretic action, reducing urine output primarily as a result of reduced glomerular filtration rate (Fig 10.2). Reduced tubular transport maximum for glucose (a measure of the maximum ability to reabsorb glucose and therefore of the functional tubular mass) also provides indirect evidence of an angiotensin II-induced reduction in the filtering population of glomeruli (Fig 10.2). This has been shown more directly by a technique which uses ferrocyanide to visualise filtering glomeruli by precipitation of filtered ferrocyanide as Prussian blue (Brown et al 1980).

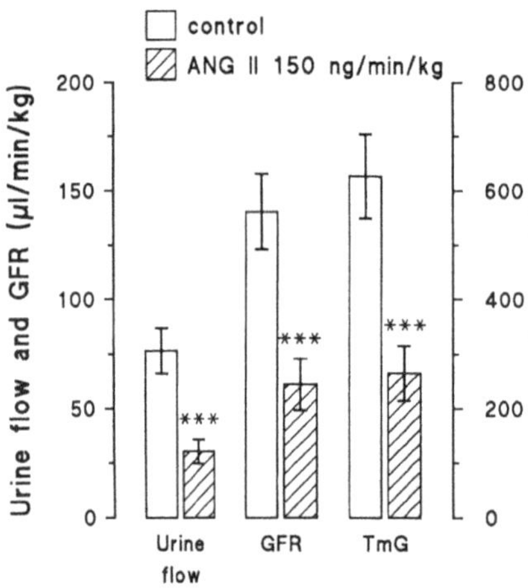

Fig. 10.2. Actions of angiotensin II infusion of rainbow trout on urine output (n=21), and glomerular filtration rates (GFR) and tubular transport maxima for glucose (TmG) (n=15). ***P<0.001, paired t test. (see Brown et al., 1980).

Angiotensin II appears to be an important component of the regulation of adaptive glomerular intermittency in teleost kidneys. Plasma angiotensin II concentrations show transitory increases in euryhaline fish,like the European eel, as they adapt from fresh water to sea water. Plasma angiotensin II concentrations rose from 8 x 10^{-12}M in fresh water to 6 x 10^{-11}M after 6 days in sea water (Balment, Warne, Tierney & Hazon, 1993).

It must be admitted that investigation of the direct renal action of circulating vasoactive peptides, like angiotensin II and AVT, in whole animals is fraught with difficulties as systemic vascular actions can result indirectly in altered renal excretion. The fish kidney does not possess the autoregulatory capacity of the tetrapod kidney and accordingly changes in systemic blood pressure can effect altered rates of glomerular filtration. In this way the apparent renal antidiuretic action of angiotensin II in teleosts can be compromised by coincident vasopressor effects (Gray & Brown, 1985), which may explain why in some experiments in the eel angiotensin II has been reported to be diuretic (Nishimura & Sawyer, 1976). The circulating angiotensin concentrations achieved in these earlier experiments have not been routinely determined, so it is not certain from these *in vivo* studies whether physiologically relevent effects were being examined. As a means of more directly investigating renal function and the responses to known concentrations of angiotensin II, an *in vitro* perfused trout kidney preparation has been employed more recently. This involves decapitation and insertion of a cannula into the dorsal aorta to perfuse the trunk with the kidney *in situ*. Other organs are removed and the *in situ* kidney is perfused from a constant pressure head. Using this approach, angiotensin II, at concentrations within the range measured in fish plasma, was shown to be antidiuretic in the trout, supporting the whole animal studies (Dunne & Rankin, 1992).

It has become clear in the last 5 years, in mammals at least, that the classical systemic RAS occurs alongside complete tissue specific or local RAS, which can potentially operate quite independently (Hackenthal, Paul, Ganten & Taugner, 1990; Wright et al., 1995). In addition to renin it appears that both renin substrate (angiotensinogen) and converting enzyme are present within many tissues (such as the kidney and brain). We have recently obtained both physiological and molecular evidence that a local renal renin-angiotensin system also occurs in the trout (Brown, Paley, Amer & Aves, 1995). Renin is known to be present in the kidney (indeed it is part of the systemic system), angiotensin converting enzyme has been demonstrated in fish kidneys including the trout (Lipke & Olson, 1988) and our own work has demonstrated the existence of the genes for angiotensinogen in the trout kidney (Brown et al 1995), thus a complete trout renal RAS is implied. Experimental studies using the trout *in situ* perfused kidney preparation (which *de facto* cannot have a circulating renin-angiotensin system) have investigated the physiological activity of this renal Renal Angiotensin System (Amer, 1995; Brown et al 1995). These experiments involved treatment of perfused kidneys with Captopril, an angiotensin converting enzyme inhibitor (to inhibit formation of angiotensin II by the *in situ* kidney, see Fig. 10.1). In preparations perfused at the 'standard' pressure head of 38cm H_2O, Captopril had no effect on renal function, but in kidneys perfused at a 'low' pressure of 25cm H_2O (which also resulted in a reduction in perfusate flow) Captopril treatment induced a glomerular diuresis. Given the glomerular antidiuretic action of angiotensin II these studies suggest that an intrarenal system is active in conditions of low flow/pressure, and can be inhibited by the converting enzyme inhibitor (Captopril).

Angiotensin II may not be the only biologically active peptide of the RAS. In mammals at least, the first amino acid of angiotensin II can be cleaved to form a smaller peptide, angiotensin III, and 'angiotensin 1-7' , or angiotensin IV (peptides 3-8) may also be formed (see Fig. 10.1) (Henderson, Brown & Balment, 1993; Wright et al., 1995). In mammals these peptides are suggested to have physiological action (Lorenz et al., 1993; Tseng et al., 1994). We have recently examined the renal action of angiotensin III in the trout using the *in situ* perfused kidney preparation and Fig 10.3 shows its antidiuretic effect. Angiotensin III at 10^{-11}M depressed urine production in each preparation, but the effect was slight (mean 22% decrease). At the higher dose of 10^{-9}M there was a 74% decline in urine output ($P<0.01$). Thus it appears that a locally active renal renin-angiotensin system may operate through both angiotensin II and angiotensin III. In some mammalian tissues angiotensin II and III have different apparent receptor preferences (Wright et al., 1995). Whether separate receptors for angiotensin II and

III exist in the trout kidney or whether angiotensin III acts on already identified renal receptors for angiotensin II (Cobb & Brown 1994) remains to be determined.

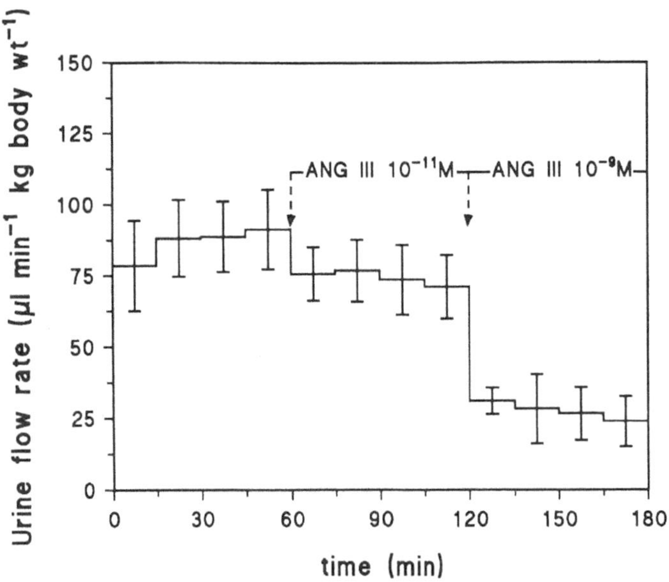

Fig. 10.3 Urine flow rates of 5 *in situ* perfused trout kidneys perfused for 1h prior to angiotensin III addition to the perfusate, initially at 10^{-11}M and after a further h at 10^{-9}M.

10.3 Arginine Vasotocin (AVT)

Arginine vasotocin (AVT), the major neurohypophysial peptide in lower vertebrates, is the homologue of mammalian arginine vasopressin or antidiuretic hormone (ADH). Although initially reported to be diuretic in teleost fish (Maetz et al., 1964), later studies in the European eel indicated that intravenous injections of low AVT doses were in fact antidiuretic, while higher doses were diuretic (Henderson & Wales, 1974; Babiker & Rankin, 1978). In rainbow trout, injected with AVT *in vivo* only antidiuretic effects were observed even at very high doses (from 40 to 550 ng/kg body weight) (Brown, 1976). None of these experiments on whole animals have been accompanied by any measurement of circulating concentrations of AVT, so the physiological relevance of these observations cannot be determined. Recently, a sensitive radioimmunoassay for AVT in fish plasma and pituitary has been developed (Warne, Hazon, Rankin & Balment, 1994). Measurement of circulating AVT concentrations in a wide range of teleost species (see Table 10.2) indicates that the physiological range of plasma AVT is 10^{-12}-10^{-11}M. Notably there was no major or consistent difference in plasma AVT levels between fresh water and sea water stenohaline teleosts. There was also no consistent difference in plasma AVT concentrations for euryhaline species when fish were fully adapted to sea water or fresh water.

Table 10.2 Plasma concentrations of AVT in freshwater and seawater stenohaline and euryhaline teleosts. Warne et al., 1994 includes some of these data.

Species		Plasma AVT (fmol/ml)	
Freshwater			
Carp (*Cyprinus carpio*)		5.7 ± 1.1	(n=9)
Brown trout (*Salmo trutta*)		2.1 ± 0.2	(n=16)
Seawater			
Plaice (*Pleuronectes platessa*)		14.4 ± 2.7	(n=10)
Dab (*Limanda limanda*)		3.7 ± 1.1	(n=9)
Saith (*Pollachius virens*)		20.2 ± 6.3	(n=7)
Cod (*Gadus morhua*)		7.2 ± 3.4	(n=5)
Euryhaline			
Flounder (*Platichthys flesus*)	[FW]	8.1 ± 0.8	(n=8)
	[SW]	8.6 ± 1.7	(n=8)
Rainbow trout (*Oncorhynchus mykiss*)	[FW]	2.7 ± 1.0	(n=3)
	[SW]	4.8 ± 1.2	(n=5)
European eel (*Anguilla anguilla*)	[FW]	0.9 ± 0.1	(n=9)
	[SW]	1.0 ± 0.1	(n=9)

Calculation of the circulating levels of AVT induced in the earlier *in vivo* studies indicates that the higher AVT doses, which were associated with a pressor action and a diuretic response, would have produced plasma AVT concentrations 2 or 3 orders of magnitude higher than the now known physiological range (Balment *et al*, 1993). Thus, in the *in vivo* studies, the described diuretic actions of AVT are likely to be only of pharmacological interest while more physiologically relevant AVT levels induce an antidiuresis.

We have recently reexamined the renal actions of concentrations of AVT within the physiological range using the *in situ* perfused trout kidney preparation (Amer & Brown 1995). This approach, using a perfusion with a constant pressure head, avoids the secondary, largely vascular effects which intravenous AVT injections can elicit *in vivo*. Administration of AVT at 10^{-9}M (probably supra-physiological) and 10^{-11}M (a physiologically relevant dose) induced a clear and immediate antidiuresis in the perfused trout kidney (Fig 10.4). The antidiuresis was more intense at the higher dose of AVT, but at both doses the effects were highly significant ($P<0.001$). Declining urine output largely reflected decreased glomerular filtration rates (Fig. 10.4), with parallel decreases in the tubular transport maxima for glucose implying reduced filtering populations of glomeruli (Fig. 10.5). More direct examination of the population of filtering glomeruli in perfused kidneys (using a ferrocyanide technique and microdissection of glomeruli: Amer & Brown, 1995) has provided direct evidence that physiological concentrations of AVT reduce the filtering population of glomeruli (Fig. 10.6).

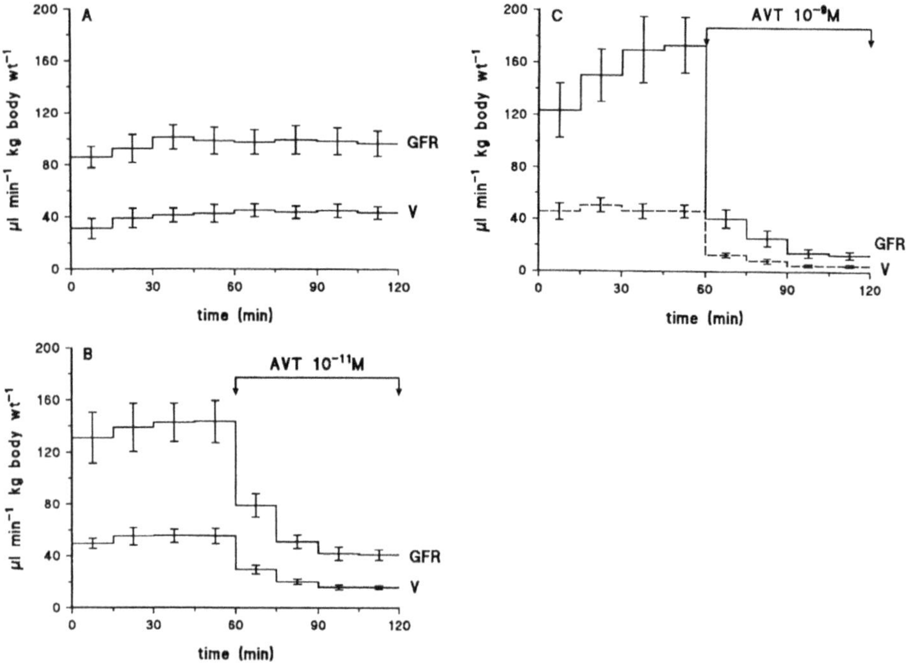

Fig 10.4. Urine flow rates (V) and glomerular filtration rates (GFR) of control (A) (n=5) and AVT-treated (B & C) *(n=6 in each group) in situ* perfused trout kidneys perfused for 2h. Experiments described and some of these data included in Amer & Brown, 1995.

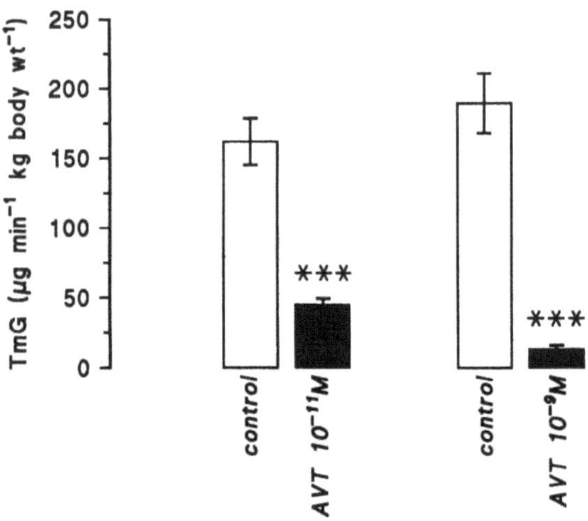

Fig. 10.5. Tubular transport maxima for glucose (TmG) of the *in situ* perfused trout kidney. The renal clearance immediately prior to AVT treatment (at 10^{-9}M or 10^{-11}M) and for the 4th collection period during treatment (n=6 per group) are shown; ***$P<0.001$, paired t test. (see Amer & Brown, 1995).

From these results it is clear that even in fresh water adapted trout, levels of AVT within the physiological range will exert antidiuretic action. Potentially, decreased circulating AVT concentrations *in vivo* may release some of this down-regulation of the filtering population of glomeruli and thereby elevate urine output in response to specific environmental/physiological conditions. Indeed, it is apparent that AVT secretion, as indicated by plasma AVT concentration, is sensitive to osmotic stimuli (Warne & Balment, 1995). In seawater adapted flounder acute hypertonic stimulii, such as the intraperitoneal (ip) injection of 1M NaCl to raise plasma osmolality by 9 mosmol/kgH_2O ($P<0.05$), resulted in a significant rise in plasma AVT concentration by comparison with isotonic saline-treated controls ($P<0.05$) (Fig 10.7). This sensitivity of AVT secretion to plasma tonicity was also evident when coincident measures of plasma AVT and sodium were analyzed in a sample of 45 sea water adapted flounders (Fig 10.8). Interestingly, this association between plasma tonicity and AVT concentration is never demonstrable in fresh water adapted fish (Balment et al., 1993), where perhaps different volume status and lower plasma osmolality and ion concentrations act as contra-indicatory stimuli. This type of interaction between volume status and plasma tonicity has been clearly demonstrated for vasopressin secretion in mammals, where also the concept of a threshold for osmotic stimulation of vasopressin secretion is well established (Robertson et al., 1976; Cowley, 1988).

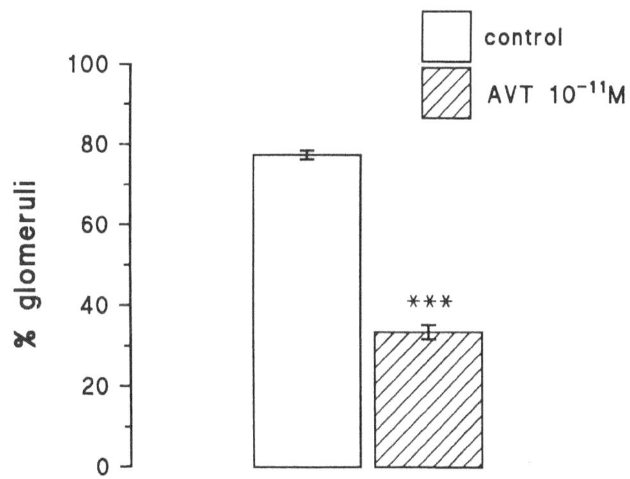

Fig. 10.6 Effect of AVT (10^{-11}M) on the population of filtering glomeruli of the *in situ* perfused trout kidney (n=6 in each group). ***$P<0.001$, Mann Whitney U test. Data derived from those presented in Amer & Brown, 1995.

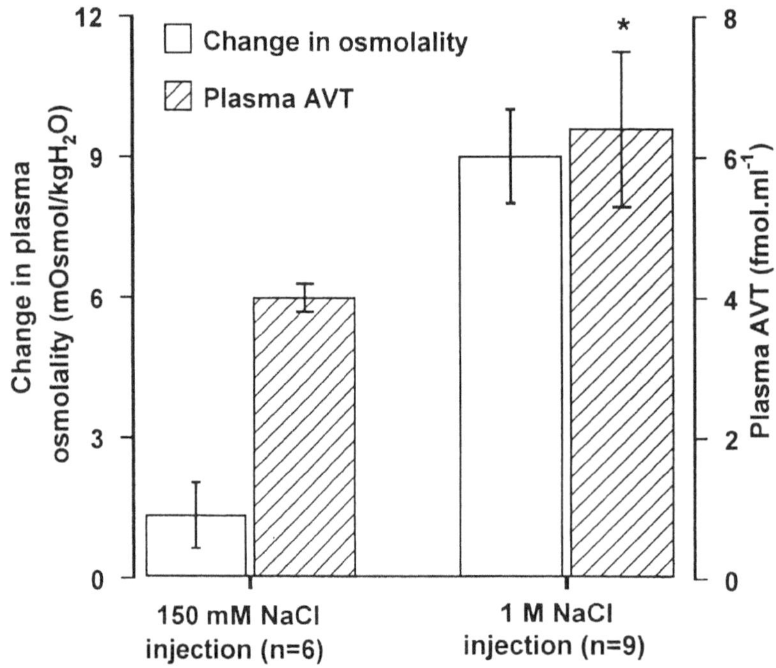

Fig 10.7 Change in blood osmolarity and associated plasma AVT concentrations 60min after ip injection of isotonic (150mM NaCl, n=6) or hypertonic (1M NaCl, n=9) saline (0.5ml/100g body wt) in sea water-adapted flounder. (see Warne & Balment, 1995)

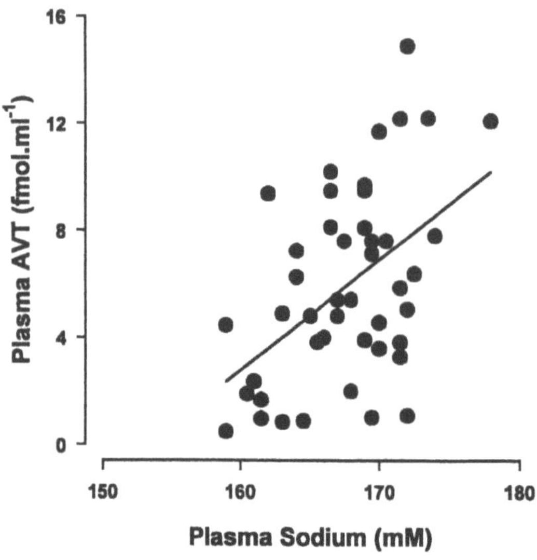

Fig. 10.8 Plasma arginine vasotocin (AVT) concentration plotted against corresponding Na^+ concentration in blood samples taken by dorsal aortic puncture from otherwise untreated seawater adapted flounder (see Warne & Balment, 1995).

As indicated above AVT has both local and systemic vasoconstrictor actions and AVT may play a role in blood pressure regulation or regional blood flow distribution (Warne & Balment, 1996). Accordingly, it might be anticipated that AVT secretion is sensitive to changes in blood pressure. Figure10.9 shows some preliminary data from studies in sea water adapted flounder in which the effects of acute hypotension induced by administration of the smooth muscle relaxant, Papaverine, has been examined (Warne & Balment, unpublished). This acute hypotensive stimulus, which we have used in previous studies (Balment & Carrick, 1985) results in a 5-7 mmHg transient (15-20 min) fall of arterial blood pressure and was consistently associated with increased plasma AVT concentration.

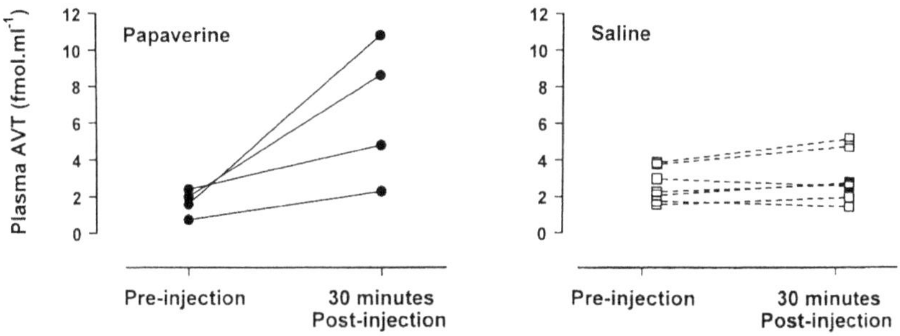

Fig. 10.9 Effect of intra-arterial injection of Papaverine (10mg/kg) on plasma AVT concentration of seawater adapted flounders. Control fish were injected with saline. Lines connect plasma AVT concentration prior to and 30 min after. Papaverine or saline injection.

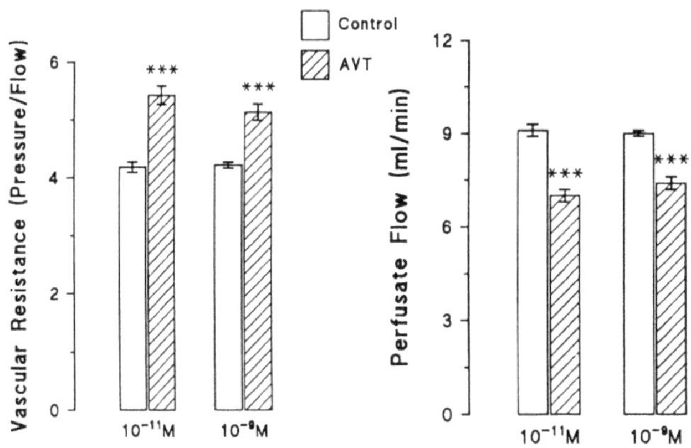

Fig. 10.10. The effect of AVT at 10^{-9}M and 10^{-11}M on measured perfusate flow and estimated vascular resistance of 6 perfused trout trunks ***P<0.001, paired t tests.

The renal antidiuretic actions of AVT appear to result largely from activation of vascular receptors and a vasoconstrictor effect to reduce glomerular rate. In the *in situ* perfused kidney and trunk preparation, AVT treatment increased vascular resistance causing a significant reduction in the measured perfusate flow (Fig. 10.10).

The vascular effects of AVT in the *in situ* perfused trout kidney seem likely to reflect the existence of V_1-type vascular receptors in the kidney but these have yet to be confirmed (Mahlmann et al., 1994). Free water clearance of the *in situ* perfused trout kidney, as a proportion of glomerular filtration rate (GFR), was significantly reduced (P<0.01) in response to 10^{-11}M AVT, decreasing from 14.5% to 12.8% . This indicates a small increase in reabsorption of filtered water by the renal tubules and thus a small but significant tubular contribution to the antidiuresis induced by a physiological dose of AVT. In tetrapods the tubular antidiuretic action of AVT or vasopressin is associated with the presence of V_2 type receptors in the distal tubule. These V_2 type receptors are coupled to adenylate cyclase by stimulatory G proteins, their occupancy by AVT resulting in a rise in intracellular cAMP levels and a consequent cascade leading to the insertion of water channels or aquaporins in the apical surface of the tubular cell. The increased tubular water permeability induced is responsible for the tubular antidiuresis. It is generally held that this type of V_2 receptor arose in the vertebrate series with the emergence of tetrapods onto land and is not present in fish (Pang, 1983). However, it appears that the trout kidney does possess receptors which recognise AVT and responds in a dose dependent manner with increased cAMP production (Perrott, Sainsbury & Balment, 1993). Figure 10.11 shows a typical response of a dispersed nephron preparation from fresh water adapted rainbow trout kidney incubated with different concentrations of AVT or forskolin. Notably, significant stimulation of cAMP accumulation was achieved at AVT concentrations as low as 10^{-12}M. This profile of response is consistent with the presence of V_2 type receptors for AVT in the trout kidney. However, it remains to be established whether these V_2 like receptors are responsible for mediating the tubular antidiuretic action of AVT observed in the perfused trout kidney. Interestingly, it became evident in these dispersed nephron studies that tissue derived from 50% sea water adapted trout were less sensitive to AVT than tissue from fresh water fish (Harding & Balment, 1995). As shown in Figure 10.12, significant stimulation of cAMP accumulation only occurred at 10^{-8}M AVT or above.

The lower AVT sensitivity of nephron preparations from 50% seawater-adapted trout may indicate that AVT receptor numbers are down-regulated during seawater adaptation in the trout. This would clearly afford adaptive changes in AVT contributions to control of renal function in the absence of any sustained alteration in circulating AVT concentration (see Table 10.2).

There are major differences in the nephron structures of fresh water and seawater teleosts. In addition to the reduced glomerular size and number in marine species the distal diluting segment, characteristic of the fresh water teleost tubule, is also often absent. It is of interest, therefore, that when the examination of AVT

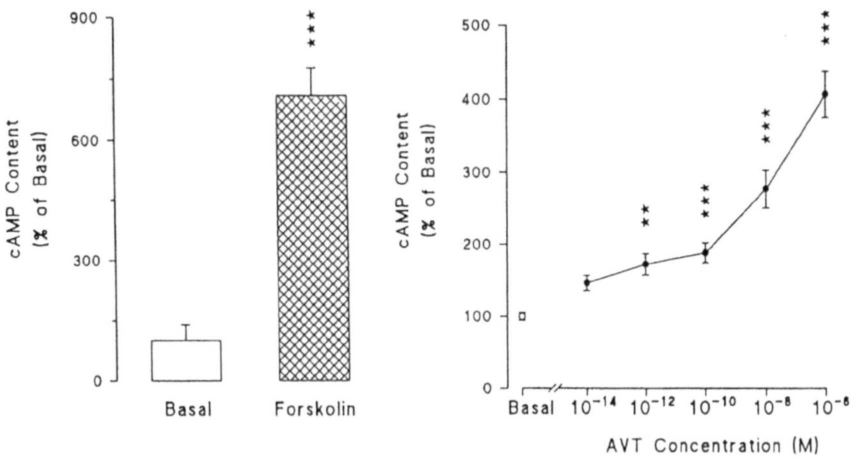

Fig. 10.11

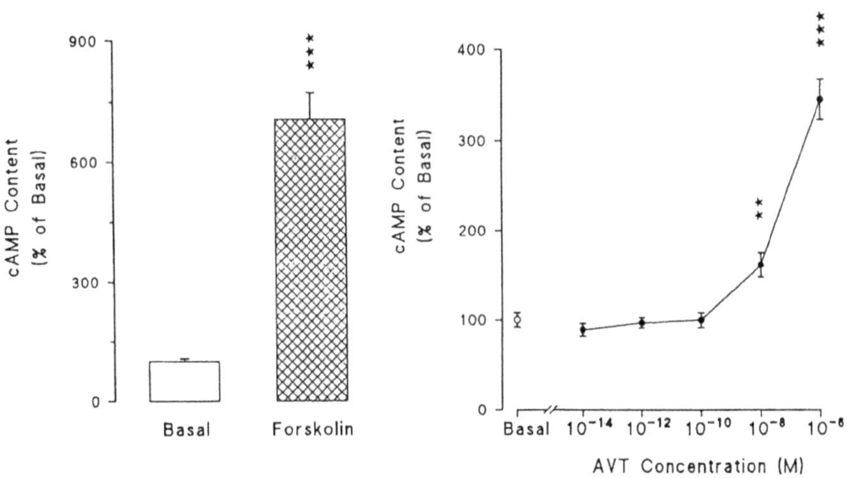

Fig. 10.12

Figs. 10.11 & 10.12 Arginine vasotocin stimulation of cAMP accumulation (% basal, non stimulated level) in typical dispersed nephron preparations of fresh water (Fig. 10.11) and 50% seawater adapted trout (Fig 10.12) (see Harding and Balment, 1995 for experimental protocol). Stimulation of cAMP accumulation by forskolin ($5x10^{-6}$M) also shown.

responsive renal adenylate cyclase was extended from trout to species of marine origin (plaice, flounder, cod) it became evident that dispersed nephrons from such species were unresponsive (Harding & Balment, 1995). This suggests that the V_2 like receptor described in trout may be located in the distal segment of the tubule.

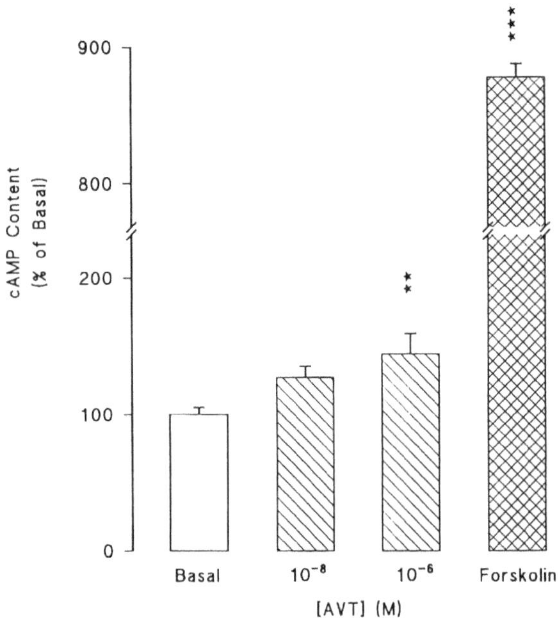

Fig 10.13 Stimulation of cAMP accumulation (as % basal, unstimulated level) by forskolin (5 x 10^{-6}M) and AVT (10^{-6}M) in bladder of seawater adapted flounder.

Embryologically the bladder is considered to be derived from the distal segment of the teleost nephron. It is perhaps significant, therefore, that when flounder bladder was examined (see Fig. 10.13) AVT was able to stimulate adenylate cyclase activity suggesting that AVT may contribute to post-renal modification of urine in this marine species (Harding & Balment unpublished observations).

As indicated in this overview there have been major advances in recent years in establishing the physiological relevance of AVT and angiotensin control of teleost renal function. In the process, some of the earlier models of hormonal input have been discarded with the realisation that many reported effects of these hormones were pharmacological artefacts produced by the very high, nonphysiological doses often employed in earlier studies. The renal actions of these hormones appear to involve adjustments in glomerular filtration rates and tubular transport. We can anticipate further rapid progress in understanding their precise mechanisms of renal action as molecular biological techniques are applied to locate specific nephron sites and mechanisms controlling expression of renal receptors for these important peptide hormones.

References

Amer, S. (1995). Endocrine and paracrine control of renal function in the *in situ* perfused trunk of the rainbow trout. PhD thesis, University of Exeter.

Amer, S. & Brown, J.A. (1995). Glomerular actions of arginine vasotocin in the in situ perfused trout kidney. Am. J. Physiol. 269, R775-780.

Babiker, M.M. & Rankin, J.C. (1978). Neurohypophysial hormone control of kidney function in the European eel (*Anguilla anguilla*) adapted to seawater or freshwater. J. Endocrinol. 78, 347-358.

Balment, R.J. & Carrick, S. (1985). Endogenous renin-angiotensin system and drinking behaviour in the flounder. Am. J. Physiol. 248, R157-R160.

Balment, R.J., Warne, J.M., Tierney, M. & Hazon, N. (1993). Arginine vasotocin and fish osmoregulation. Fish Physiol. Biochem. 11, 189-194.

Brown, J.A. (1976). Development and application of methods for the study of renal function in lower vertebrates. PhD thesis, University of Sheffield. UK.

Brown, J.A., Oliver, J.A., Henderson, I.W. and Jackson, B.A. (1980). Angiotensin and single nephron glomerular function in the trout, *Salmo gairdneri*. Am J. Physiol. 239, R509-514.

Brown, J.A., Paley, R.K., Amer, S. & Aves, S.J. (1995). Evidence for an intrarenal renin-angiotensin system in the rainbow trout. J Endocrinol. 147 S, P79.

Cobb, C.S. & Brown, J.A. (1994). Characterisation of angiotenin II binding to glomeruli from rainbow trout *Oncorhynchus mykiss* adapted to fresh water and seawater. Gen. Comp. Endocrinol. 94: 104-112.

Cowley, A.W.Jr. (1988). Vasopressin and blood pressure regulation. Clin Physiol. Biochem 6, 150-162.

Duff, D.W. & Olson, K.R. (1986). Trout vascular and renal responses to atrial natriuretic factor and heart extract. Am.J. Physiol. 251, R639-R642.

Dunne, J.B. & Rankin, J.C. (1992). Effects of natriuretic peptide and AII on salt and water excretion by the perfused rainbow trout kidney. J. Physiol., 446, p92.

Gray, J.C. & Brown, J.A. (1985). Renal and cardiovascular effects of angiotensin II in the rainbow trout, *Salmo gairdneri*. Gen Comp Endocrinol 59, 375-381.

Hackenthal, E., Paul, M., Ganten, D., & Taugner, R. (1990). Morphology, physiology and molecular biology of renin secretion. Physiological Reviews 70. 1067-1116.

Harding, K.E. & Balment, R.J. (1995). Effects of AVT on cAMP accumulation in renal tubules of teleost fish. J. Endocrinol. 147, Suppl. P60.

Henderson, I.W., Brown, J.A & Balment, R.J. (1993). The renin-angiotensin system and volume homeostasis. In: New Insights into Vertebrate Kidney Function (Eds. A. Brown, R.J. Balment & J.C. Rankin). pp31-349, Cambridge University Press, Cambridge.

Henderson, I.W. & Wales, N.A.M. (1974). Renal diuresis and antidiuresis after injections of arginine vasotocin in the freshwater eel (*Anguilla anguilla* L). J. Endocrinol. 61, 487-500.

Kaiya, H. & Takei, Y. (1996). Plasma levels of atrial and ventricular natriuretic peptide in freshwater- and seawater-adapted eels. Gen. Comp. Endocrinol. (In Press).

Lee, J. & Malvin, R.L. (1987). Natriuretic response to homologous heart extract in aglomerular toadfish. Am. J. Physiol. 252, R1055-R1058.

Lipke, D.W. & Olson, K.R. (1988). Distribution of angiotensin-converting enzyme like activity in vertebrate tissues. Physiol. Zool. 61, 420-428.

Lorenz, J.N., Wehprecht, H., He., X.R., Scott, O., Briggs, J.P. & Schnermann, J. (1993). Effects of adenosine and angiotensin on macula densa-stimulated renin secretion. Am .J. Physiol. 265, F187-194.

Maetz, J., Bourget, J., Lahlou, B. & Hourdry, J. (1964). Peptides neurohypophysaires et osmoregulation chez *Carassius auratus*. Gen. Comp. Endocrinol. 4, 508-522.

Mahlmann, S., Meyerhof, W., Hausmann, H., Heierhorst, J., Schronrock, C., Zwiers, H., Lederis, K. & Richter, D. (1994). Structure. function and phylogeny of [Arg^8]vasotocin receptors from teleost fish and toad. Proc. Natl. Acad. Sci. USA 91, 1342-1345.

Nishimura, H. & Sawyer, W.H. (1976). Vasopressor, diuretic and natriuretic responses to angiotensin by the American eel, *Anguilla anguilla rostrata*. Gen. Comp. Endocrinol. 29, 337-348.

Pang, P.K.T. (1983). Evolution of control of epithelial transport in vertebrates. J.Exp. Biol. 106, 283-299.

Perrott, M.N., Sainsbury, R.J. & Balment, R.J. (1993). Peptide hormone-stimulated second messenger production in the teleostean nephron. Gen. Comp. Endocrinol. 89, 387-395.

Rankin, J.C., Henderson, I.W. & Brown, J.A. (1983). Osmoregulation and the control of kidney function. In: 'Control Processes in Fish Physiology' (Eds J.C. Rankin, T.J.Pitcher & R. Duggan) pp 66-88.Croom Helm, London.

Robertson, G.L., Shelton, R.L. & Athar, S. (1976). The osmoregulation of vasopressin. Kidney Internat. 10, 25-37.

Takei, Y (1994). Structure and function of natriuretic peptides in vertebrates. Perspectives in Comparative Endocrinology, National Research Council of Canada, pp149-154.

Takei, Y. & Balment, R.J. (1993). Biochemistry and physiology of a family of eel natriuretic peptides. Fish Physiol. Biochem. 11, 183-188.

Tseng, C.J., Chou, L.L., Ger, L.P. & Tung, C.S. (1994). Cardiovascular effects of angiotensin III in brain-stem nuclei of normotensive and hypertensive rats. J Pharmacol & Exp. Therapeut. 268, 558-564.

Warne, J.M. & Balment, R.J. (1995). Effect of acute manipulation of blood volume and osmolality on plasma [AVT] in seawater flounder. Am.J. Physiol. 269, R1107-R1112.

Warne, J.M. & Balment, R.J. (1996). Vascular actions of neurohypophysial peptides in the flounder. Fish Physiol. Biochem. (In Press).

Warne, J.M., Hazon, N., Rankin, J.C. & Balment, R.J. (1994). A radioimmunoassay for the determination of arginine vasotocin (AVT): plasma and pituitary concentrations in fresh- and seawater fish. Gen. Comp. Endocrinol. 96, 438-444.

Wright, J.W., Krebs, L.T. Stobb, J.W. & Harding, J.W. (1995). The angiotensin IV system- functional implications. Front. in Neuroendocrinol. 16, 23-52.

Arginine vasotocin (AVT) controls renal sodium and water excretion in birds through interaction with a new ADH receptor subtype

R. Gerstberger

Max-Planck Institute for Physiological and Clinical Research, W.G. Kerckhoff – Institute, Parkstrasse 1, D-61231 Bad Nauheim, Germany

11.1 Body fluid homeostasis

The body-intrinsic, extracellular fluid (ECF) represents the compartment responsible for the controlled exchange between the intracellular milieu and the external environment in birds as all other classes of vertebrates. With regard to ionic composition, sodium (Na^+) and chloride (Cl^-) contribute preferentially to the osmotic pressure of the ECF. Maintenance of the sodium concentration (tonicity) and volume of the ECF under varying external conditions are a necessary prerequisite for all cellular and integrative functions, and a homeostatic control circuit has evolved to control these ECF parameters (Fig. 11.1.). Minute alterations in either ECF tonicity or volume are perceived by systemic osmo- und volume receptive elements as well as central nervous osmo- and sodium sensors, well characterized in mammals and birds. All these afferent signals are finally

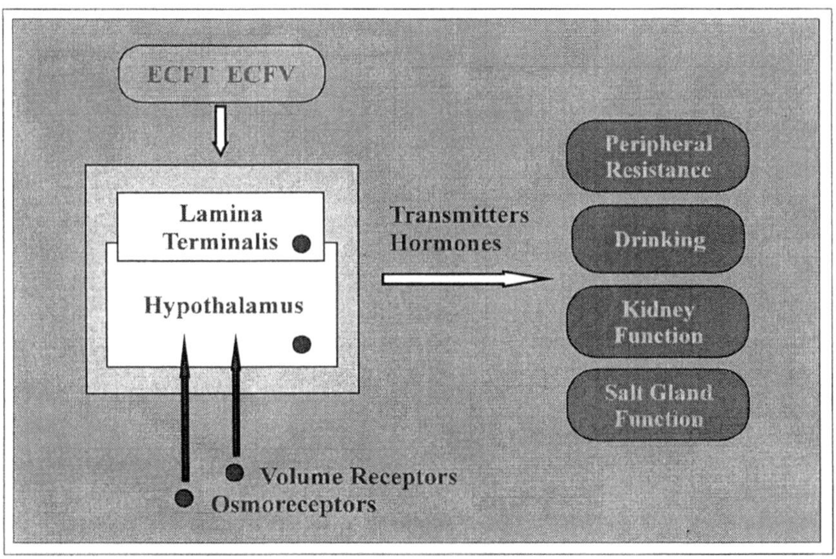

Fig. 11.1. Homeostatic control circuit to maintain body fluid homeostasis in birds

conveyed to the hypothalamus of the diencephalon, where complex integration takes place in mammals (Andersson, 1978; Bie, 1980) or birds (Gerstberger, 1996; Simon, 1982). Besides a centrally mediated alteration in the drive to drink water, peripheral osmoregulatory target organs such as the kidneys, the cardiovascular system or supraorbital salt secreting glands in the case of marine birds are subsequently activated or inhibited via the autonomic nervous system and (neuro-) hormones to reestablish ECF tonicity or volume in avian species (Fig. 11.1.).

The kidneys represent the main organ system in birds to maintain body fluid homeostasis and are composed of multiple lobules as functional units, each of them showing a cortical and a medullary component, forming a , "mushroom"-like body (Fig. 11.2.). In addition to the arterial blood reaching each lobule via an intralobular artery, venous blood derived from the mesenteric and iliac veins feeds a complicated portal venous system supplying the cortical portion of each lobule. Different from the situation in mammals on one hand (production of hyperosmotic urine), and reptiles on the other hand (production of hypo- to isosmotic urine), birds can concentrate their urine to 600-700 mOsm/kg. This fact is due to the presence of both, "mammalian-type" (M) and "reptilian-type" (R) nephrons (Dantzler, 1989). M-type nephrons possess large glomeruli (diameters of 80-110 µm) and a loop of Henle. Being therefore able to concentrate urine above plasma

KIDNEY ANATOMY

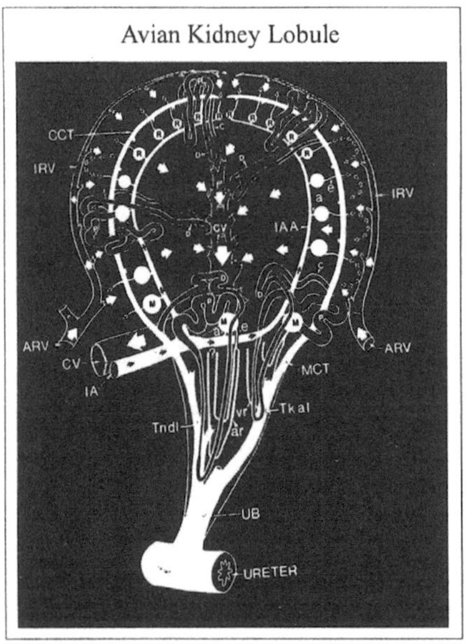

Fig. 11.2. Each lobule is supplied by arterial blood via the intralobular artery (IA) and venous blood via the afferent portal vein (ARV). Looped nephrons (M) with their loops of Henle extend into the medullary cone (Tndl, Tkal). Loop-less nephrons (R) in the cortical portion of the lobule lack a loop of Henle. The medullary collecting tubules (MCT) represent the final. segments for both nephron types.
Courtesy to Dr. R. F. Wideman, Penn. State Univ., USA (see also: Wideman,1988).

osmolality, these M-types nephrons are situated so that their loops of Henle project into the medullary cones. R-type nephrons have small glomeruli (diameters of 25-35 μm) and lack the loop of Henle. They can be found in the cortical section of each lobule and are oriented perpendicularly to intralobular arteries and cortical collecing tubules (Fig. 11.2.) (Morild et al., 1985; Wideman 1988).

11.2 The antidiuretic hormone AVT

Following central nervous integration of afferent osmo- or volume regulatory signals within the avian autonomic control circuit for the maintenance of body fluid homeostasis, altered release of the neuroendocrine antidiuretic hormone (ADH) represents one of the major routes to readjust ECF tonicity or volume. With arginine vasopressin (AVP) as the mammalian-specific ADH, arginine vasotocin (AVT) represents the avian, reptilian, amphibian and partially piscine neuroendocrine, antidiuretic principle (Dantzler, 1980, 1989). The cyclic nonapeptide (as circulating hormone) is primarily synthesized in magnocellular neurons of the paraventricluar (PVN) and supraoptic (SON) nuclei in the avian

HYPOTHALAMIC AVT

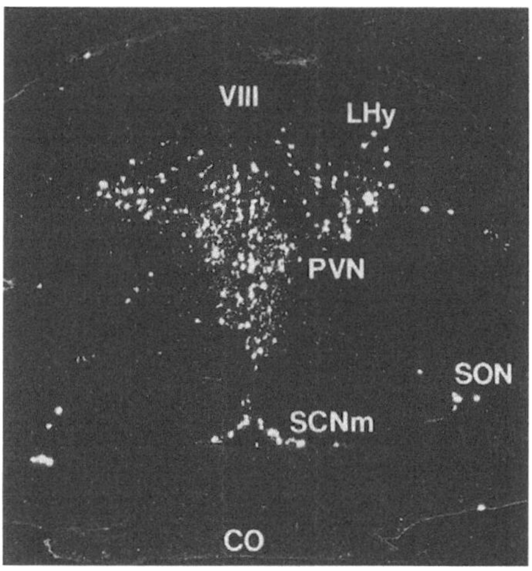

Fig. 11.3 *In situ* hybridization for AVT mRNA in the chicken hypothalamus. Both magno- and parvocellular neurons expressing AVT could be shown in the paraventricular nucleus (PVN) near the third cerebral ventricle (VIII), the supraoptic (SON) and the suprachiasmatic nucleus (SCNm) near the optic chiasm (CO). Single cells expressing AVT could be found within the lateral hypothalamus (Lhy).
Courtesy to Dr. S. Barth, FAL Celle, FRG (see also: Jurkevich et al., 1996).

hypopthalamus, with both nuclei bordering the third cerebral ventricle, as shown by immunocytochemical techniques, and recently for the chicken, quail and duck also using *in situ* hybridization (Fig. 11.3.) The amino acid sequence for the chicken vasotocin precursor peptide, derived from cDNA clones, indicated a structural organization comparable to that described for the AVT/AVP gene family from other species, with the only difference, that the C-terminal glycopeptide moiety appears not to be cleaved off from the neurophysin part (Hamann et al., 1992). Released from the axons of the magnocellular neurons terminating within the median eminence or the neurohypophysis, AVT circulates as free hormone in the blood at coneentrations of 5-30 pg/ml in various birds with the short biological half-life of 1-2 min (Gray & Simon, 1983; Gray & Erasmus, 1988; Stallone & Braun, 1986). Being the major osmoregulatory hormone in birds besides angiotensin II and the atrial natriuretic factor (Schutz et al., 1992; Simon et al., 1992, Wilson 1989), AVT proved to be mainly under osmotic rather than volume control in avian species. Thus, increasing plasma osmolality due to altered sodium concentration during water deprivation primarily, led to a linearily correlated rise in plasma AVT with a sensitivity of 0.3-0.4 pg/ml per mOsm/kg in

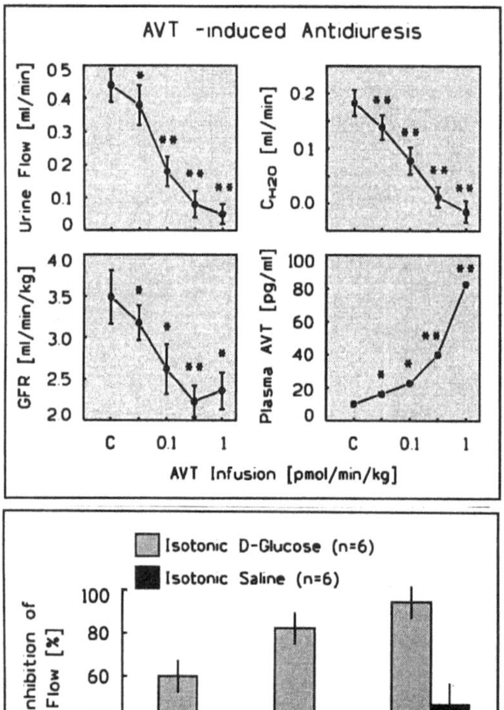

Fig. 11..4. Dose-dependent action of AVT on urine flow, GFR and free water clearance in the kelp gull (*Larus dominicanus*). Systemic intusions of AVT ranging from 0.03 - 1.0 ng/min/kg bw resulted in plasma AVT concentrations of 10 - 80 pg/ml. Courtesy to Dr. D.A. Gray, Univ. Witswatersrand, Johannesburg, South Africa (see also: Gray & Erasmus,1988)

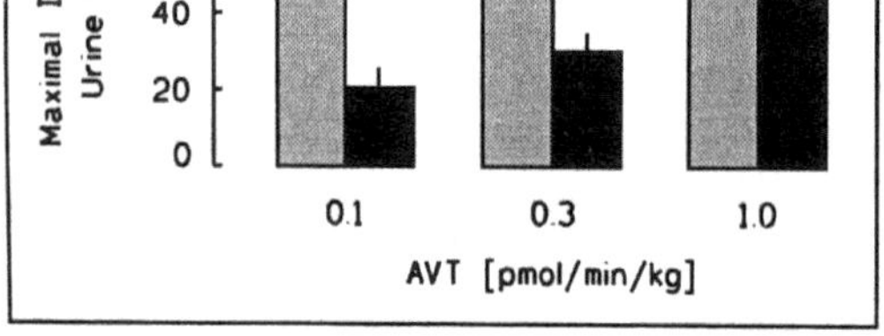

Fig. 11..5. The antidiuretic action of AVT in the fresh- water-acclimated Pekin duck can be modulated by changing extracellular sodium concentration. Isotonic saline loading diminished the antidiuretic potency of AVT as compared to isosmotic D-glucose loading.

the duck (Gray & Simon, 1983) and even 1.3 pg/ml per mOsm/kg in the dehydrated chicken (Stallone & Braun, 1986). Superimposed hypo- or hypervolemia resulted in a modest alteration of the slope relating ECF tonicity and plasma AVT. Using molecular biology techniques, Mühlbauer and coworkers (1992) reported increased expression of the AVT gene in the chicken hypothalamus already 19 h after onset of water withdrawal, with plasma osmolality still showing near-basal values. Surprisingly, in chicken challenged by hypertonic saline drinking for 5 days, a dissociation between plasma osmolality and AVT plasma concentration was visible: extremely high plasma osmolalities were accompanied by only moderate plasma AVT concentrations. Hypothalamic mRNA levels were well below AVT mRNA levels with 19 h water deprivation. Investigating the interesting question, at which point during embryogenesis the hypothalamic AVT system in birds matures, and when there is clear osmoresponsiveness of the system established, Klempt and coworkers (1992) could show appearance of plasma AVT at embryonic day E16 (out of 21 days *in ovo*). Osmotic stimulation in embryos at various stages demonstrated an increase in plasma AVT related to hypertonicity of the plasma at embryonic day E18, suggesting maturation of the osmotically driven AVT system in birds already before hatching.

11.3 AVT-induced antidiuresis and antinatriuresis

Regulatory functions of the hypothalamo-neurohypophyseal ADH system with regard to renal salt and water elimination have been described for all classes of vertebrates (Braun & Dantzler, 1974; Cree, 1988; Dantzler, 1980,1989; Pang et al., 1982;) with the most detailed knowledge available concerning actions of mammalian-specific AVP. In addition to its well known modulation of water permeability in the collecting duct of the mammalian kidney, AVP has been proposed to influence glomerular filtration, renal medullary blood flow or sodium transport in the ascending loop of Henle (Bankir et al., 1989; Kirk & Schafer, 1992). ADH-induced renal actions in non-mammalian vertebrates have been especially tested in amphibian and avian, and to a lesser extent fish and reptilian species, using the endogenous hormone arginine vasotocin (AVT) (Dantzler, 1989). Physiological glomerular actions of AVT have been described for the rainbow trout (Amer & Brown, 1995). In a perfused kidney preparation, AVT at the low dose of 10^{-11} M reduced urine flow rate, GFR and free water clearance, possibly by diminishing the number of filtering glomeruli. In both anuran and urodele amphibians, AVT induced antidiuresis via action on the glomerular vasculature (Boyd & Moore, 1990; Cree, 1988; Pang et al., 1982). In early studies on the unanesthetized water snake *Nerodia sipedon*, bolus injections of AVT as low as 3.1ng/kg caused reduction in GFR and urine flow at unchanged arterial pressure (Dantzler 1967).

As far as avian species are concerned, systemically administered AVT reduced renal water elimination in the Pekin duck, Kelp gull and the domestic fowl (Gerstberger et al., 1985; Stallone and Braun, 1985; Gray and Erasmus, 1988). Antidiuresis, antinatriuresis despite augmented urinary sodium concentration, reductions in glomerular filtration rate and renal plasma flow could be observed in the gull and duck after systemic infusion of AVT in physiologically relevant concentrations using tracer dilution techniques, thus indicating both glomerular and tubular actions of the neurohormone (Fig. 11.4.) (Gerstberger et al., 1985; Gray & Erasmus, 1988). To elucidate the role of circulating AVT for the control of kidney function in the white Leghorn chicken, AVT (1 ng/min/kg bw) was applied to conscious animals under conditions of steady-state diuresis driven by systemic infusion of hypotonic NaCl, conditions comparable to those employed in studies in ducks (Gerstberger et al., 1994) and gulls (Gray & Erasmus, 1988). AVT infusion markedly reduced urine flow rate and slightly enhanced natriuresis, quite different from experiments in the duck and gull, where antidiuresis was accompanied by pronounced antinatriuresis. GFR was only decreased by some seven percent (Fig. 11.6.) (Brummermann & Braun, 1995). Interestingly, in white Leghorn chicken with diabetes insipidus (DI), AVT application induced a strongly stimulated natriuresis, at least partially due to augmented GFR, at only slightly reduced urine flow rate (Fig. 11.6.) (Brummermann & Braun, 1995). The reduction in the potency of AVT to induce antidiuresis is possibly as a consequence of AVT receptor down-regulation in tubular components of the chicken kidney, whereas an explanation for the markedly enhanced natriuresis remains to be investigated in future studies.

DIABETES INSIPIDUS IN BIRDS

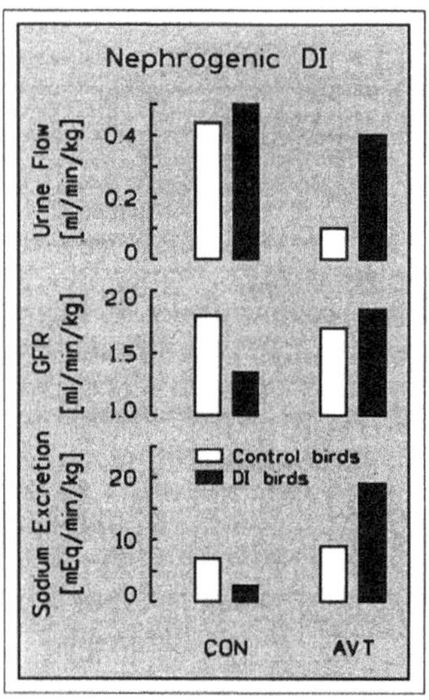

Fig. 11.6. Alterations in urine flow, glomerular filtration (GFR) and renal sodium excretion in white Leghorn chicken of control or Diabetes insipidus (DI) strain due to systemic infusion of AVT for 70 min (1 ng/min/kg bw). Constant diuresis was induced by continuous, intravenous infusion of 0.4 ml/min 50 mM NaCl. AVT infusion did not alter mean arterial pressure or heart rate in these animals.
Courtesy to Dr. E.J. Braun, Univ. Arizona, Tucson, USA (see also: Brummermann & Braun 1995)

In the freshwater-acclimated duck, the antidiuretic activity of exogenously applied AVT in physiological concentrations showed a clear dependence on the sodium status of the ECF. Thus, AVT proved to be more potent during infusion of isosmotic, not hypotonic D-glucose infusion, as compared to infusion of

AVT AND RENAL SODIUM AND WATER EXCRETION: PHARMACOLOGY

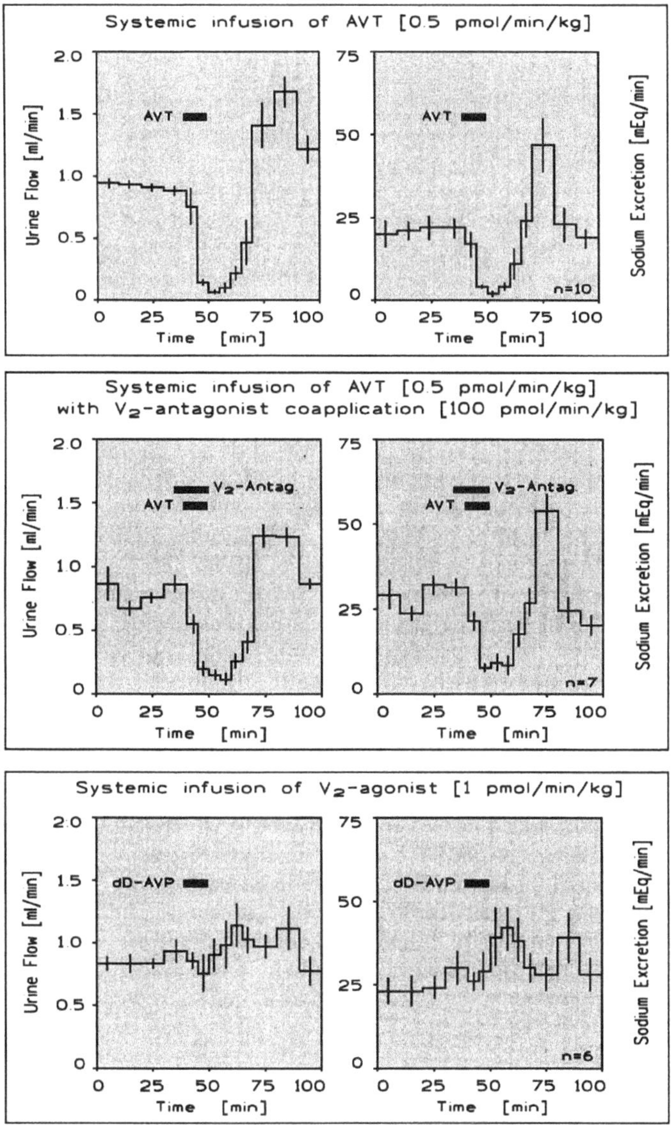

Fig. 11.7. Time-courses of urine flow rate (UF) and sodium excretion (Na-Exc) in freshwater-acclimated ducks during steady-state diuresis driven by continuous intravenous infusion of 0.7 ml/min of 200 mOsm/kg D-glucose solution and 0.3 ml/min of 200 mOsm/kg NaCl solution. Pharmacological manipulations to physiologically characterize the renal ADH receptor were carried out by adding AVT alone to the infusion (TOP PANEL), AVT together with a 200-fold higher dose of a specific V_2-receptor antagonist (MIDDLE PANEL) or by V_2-agonist dDAVP alone (BOTTOM PANEL)

isosmotic NaCl infusion (Fig. 11.5). Interaction of intraluminal sodium concentration with ADH activity is also discussed for mammals (Kirk & Schafer, 1992).

Taking the complex bird-specific renal cytoarchitecture of both looped and loopless nephrons into accoutn (Morild *et al.*, 1985: Wideman, 1988), single-nephron glomerular filtration rates as determined in the starling and quail, support physiological targeting of renal glomeruli belonging to loopless nephrons by AVT (Braun & Dantzler, 1974; Braun, 1991). The pioneering work of Braun and Dantzler on AVT-induced avian antidiuresis (1974) clearly showed the exclusive reduction of filtering nephrons lacking a loop of Henle during systemic application of AVT in the desert quail. With regard to possible additional action of AVT on M-type nephrons, transepithelial voltage as well as chloride transport determined in isolated thick limbs of Henle (TAL) of the quail *Coturnix coturnix* was not modified by AVT, and the neuropeptide did not increase the osmotic water permeability of that segment in the looped nephrons of the avian kidney, supporting possible actions of AVT downstream of the TAL (Nishimura et al., 1984).

To date, studies concerned with the characterization of the receptor system in the avian kidney involved in the AVT-induced antidiuretic and antinatriuretic action have not been performed in detail. The Pekin duck was therefore chosen as a well characterized model system to elucidate the avian renal ADH receptor system. Physiological experiments were performed in conscious ducks (n=24) to evaluate the (anti-)diuretic and (anti-)natriuretic actions of AVT, and for comparison also mesotocin, the second nonapeptide system of the hypothalamo-neurohypophyseal axis in birds (Gerstberger et al., 1985). Ducks suspended in a cotton sling allowing free movement of the neck and the legs received a continuous intravenous infusion of 0.7 ml/min of 200 mOsm/kg D-glucose and 0.3 ml/min of 200 mOsm/kg NaCI solutions. During the subsequently established steady-state diuresis, renal clearances of water, sodium and chloride amounted to 95, 65 and 58 percent, respectively, at stable plasma concentrations of AVT, angiotensin II and atrial natriuretic factor as important osmoregulatory hormones. Bird-specific AVT infused systemically for 10 min during steady-state diuresis in concentrations of 0.1, 0.5, 1.0 and 5.0 ng/min/kg bw resulted in plasma AVT concentrations of 5.5+0.6, 11.8+2.1, 18.5+3.1 and 58.0+8.8 pg/ml, respectively, as compared to basal values of 4.3+0.5 pg/ml (determined by specific RIA according to Gray & Simon, 1983). Urine flow rate was dose-dependently reduced at concomitantly elevated urine osmolality due mainly to enhanced concentrations of sodium and chloride. Both absolute and time-integrated values of antidiuresis revealed saturation kinetics for hormone action, with 1.0 ng/min/kg bw representing the AVT concentration almost causing anuria. With regard to osmolal excretion, antinatriuresis proved to be already maximal at 0.5 ng/min/kg bw concerning both absolute and time-integrated values (Fig. 11.7.). Mammalian-specific AVP infused at 0.5, 1.0 and 5.0 ng/min/kg bw induced comparable effects with regard to renal water and sodium elimination at, however, markedly reduced activity (not shown). Thus, 5 ng/min/kg bw AVP exhibited renal antidiuresis equipotent to 0.5 ng/min/kg bw AVT, at even weaker antinatriuretic activity.

To further elucidate the participation of classical ADH receptors of the V_2-subtype in ADH-induced antidiuresis, the V_2-specific agonist dDAVP was administered systemically to conscious diuretic ducks in concentrations of 1, 10 and 100 ng/min/kg b.w.. At 1 ng/min/kg b.w., comparable to the 0.5 ng/min/kg b.w. for AVT, the V_2 -agonist caused neither antidiuresis nor antinatriuresis in the duck (Fig. 11.7.), but rather a slightly delayed diuresis and natriuresis. Only at concentrations of 10 and 100 ng/min/kg bw did dDAVP induce a significant increase in urinary osmolality and sodium concentration. Urine flow rate was reduced to half-maximal values of 0.4-0.5 ml/min as compared to 0.9 ml/min during control steady-state diuresis, but never fully suppressed. With regard to time-integrated antidiuresis and antinatriuresis, dDAVP did not alter the elimination of sodium and chloride, but led to a partial retention of water. To further elucidate the possible involvement of a V_2-like or V_1-like ADH receptor subtype in AVT-induced antidiuresis, various V-$_1$ V_2- and oxytocin-specific antagonists were infused at the high concentration of 100 ng/min/kg bw systemically prior and during the application of AVT (0.5 ng/min/kg bw). Amongst the specific antagonists tested were [d(CH_2)$_5$Sar^7,Arg^8]VP and [d(CH_2)$_5$,Tyr^2 (Me),Arg^8]VP for the V_1-subtype, and [Adamantaneacetyl,D-Tyr^2(Et),Val^4,Abu^6, Arg^{8,9}]VP and [d(CH_2)$_5$,D-Phe^2,Ile^4,Ala^9-NH_2]AVP for the V_2-subtype (Fahrenholz et al., 1984; Gerstberger et al., 1994). Having no agonistic potency themselves, as tested beforehand, neither V_2- nor V_1,-antagonists inhibited the marked antidiuresis and -natriuresis induced by the endogenous hormone AVT (Fig. 11.7.). These physiological data clearly demonstrate the expression of an ADH-receptor subtype in the avian kidney which does not resemble the classical V_1- or V_2-receptor subtypes described for mammals. To prove possible cross-reactivities of AVT, AVP or the V_2-agonist dDAVP with putative mesotocin receptors, the latter peptide was applied to conscious ducks in concentrations of 1, 10 and 100 ng/min/kg bw, leading to a dose-dependent increase in urine flow rate with remarkably delayed onset and long duration of action, quite in opposite to AVT-induced antidiuresis. In addition, osmolality, sodium and chloride concentrations of the urine were augmented immediately after beginning of peptide application and decreased to basal values at the time of diuresis onset, indicating direct tubular natriuretic actions of mesotocin in the duck. The physiological pattern of renal mesotocin actions therefore appeared to be different from AVT/AVP effects with marked diuretic and natriuretic as compared to antidiuretic and antinatriuretic responses. Applying graded doses of oxytocin, the mammalian mesotocin analogue, systemically to conscious rats, renal sodium excretion proved to be linearly correlated to plasma oxytocin concentrations within physiological limits. In the presence of an oxytocin-, but not a V_1-specific receptor antagonist, natriuresis was inhibited (Verbalis et al., 1991).

11.4 The renal AVT/AVP receptor in birds

The physiological experiments performed in the duck, chicken and gull indicated both glomerular and tubular actions of the antidiuretic hormone AVT in birds to bring about antidiuresis and, in most cases, antinatriuresis. To localize putative ADH-specific binding sites in the duck kidney, receptor autoradiography was performed using [^{3}H]AVP as radioligand. Cryostat sections of duck kidneys were incubated with 10 nM of the radioligand for 90 min at RT as described before (Gerstberger & Fahrenholz, 1989; Schutz & Gerstberger, 1992). Labelling was confined to the glomeruli of loopless nephrons, areas containing distal tubules surrounding the central veins and distinct patches of medullary cone tissue (collecting duct zone), but not proximal tubules or ureteral branches of the duck kidney (Fig. 11.8.). Glomeruli of the M-type nephrons did not reveal positive staining with the radioligand. For all nephronal entities shown, specificity of the labelling was indicated by full displacement of [^{3}H]AVP using high concentrations (10^{-6} M) of either unlabelled AVP or AVT. In amphibian kidneys, autoradiographical studies revealed [^{3}H]AVP-specific binding sites within the glomerular compartment (Boyd & Moore, 1990; Kloas & Hanke, 1992). With the glomeruli representing the exclusive component stained in the amphibian kidney,

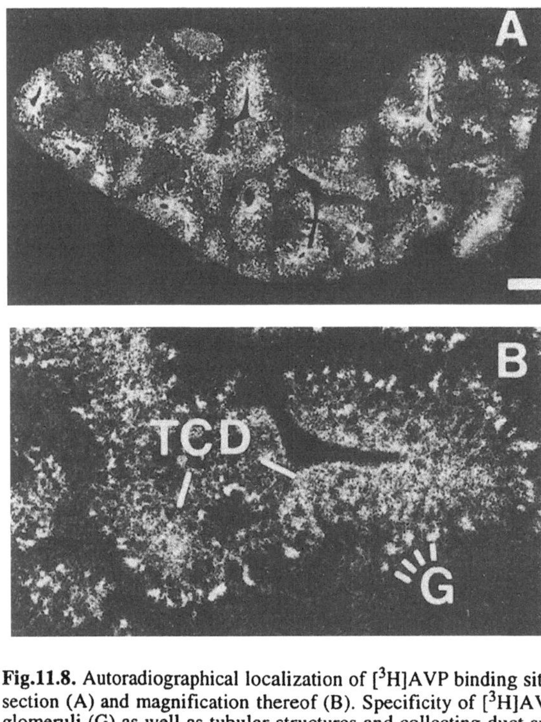

Fig.11.8. Autoradiographical localization of [^{3}H]AVP binding sites in whole duck kidney transversial section (A) and magnification thereof (B). Specificity of [^{3}H]AVP binding in loopless reptilian type glomeruli (G) as well as tubular structures and collecting duct system (TCD) was proven in absence and presence of 10^{-6} M unlabelled AVT. Bars represent 100 µm (A) and 500 µm (B), respectively.

and both inner and outer medullary structures, possibly proximal tubules and mesangial cells labelled in the rat kidney (Ammar et al., 1992; Jard et al., 1987; Jung & Endou, 1991), the ADH-receptive system of birds might represent an evolutionary state of transition from pure glomerular to medullary tubular function.

RECEPTOR BINDING

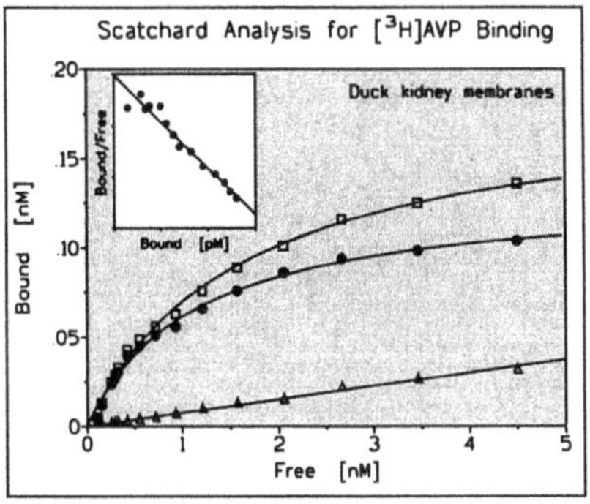

Fig.11.9. Saturation binding of [^{3}H]AVP to an enriched membrane fraction of duck nephrons in the absence (□) (total binding) and presence (Δ) (non- specific binding) of 10^{-6} M unlabeled AVT; specific binding (●) is calculated as the difference of TO and NSB. Inset shows the corresponding Scatchard analysis.

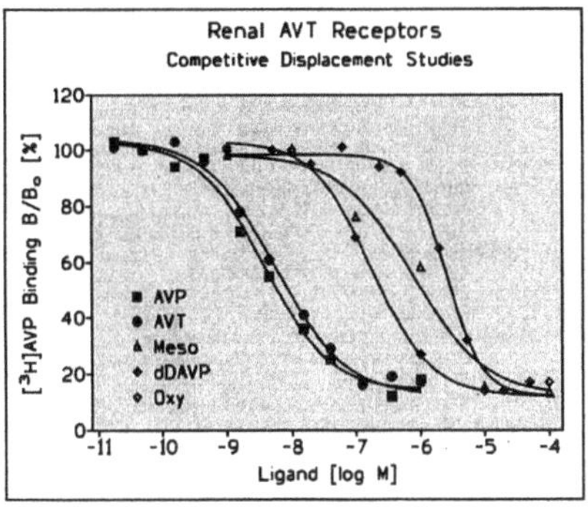

Fig.11.10. Characterization of [^{3}H]AVP binding to an enriched membrane fraction of FW duck nephrons by displacing the radioligand (2 nM) with increasing concentrations of unlabeled AVT, AVP, the V_2-specific agonist dDAVP, oxytocin or mesotocin. Various V_1- and V_2-specific antagonists did not show any potency to compete with the radio- ligand (not shown).

Scatchard analysis using an enriched plasma membrane fraction (Percoll density gradient) of duck whole kidney preparations or isolated glomeruli were performed to prove saturation of ligand binding and to determine receptor number (BmaX) and binding affinity (KD) Fig. 11.9.). With Hill coefficients not different from one, a single class of binding sites could be determined for both tissue fractions. Regulation of peptide hormone receptors during various physiological conditions with regard to either number of binding sites per cell or binding affinity for the respective hormone supports functionality of a hormone-specific "binding site". With regard to the role of ADH in body fluid homeostasis, both the plasma concentration of AVT and the ADH receptor density and affinity in the kidney were therefore determined under conditions of euhydration, dehydration and chronic salt acclimation in the duck. Short- and long-term induced elevations in fluid (ECF) tonicity induced an enhanced release of neurohypophyseal AVT and a concomitant decrease in the number of [^{3}H]AVP-specific binding sites in the kidney at unaltered binding affinity, with more pronounced effects after 48 h of water withdrawal (Fig. 11.11.). A comparable down-regulation of ADH binding sites was reported for the basolateral membranes of the rat kidney collecting tubules after 72 h of dehydration (Steiner & Phillips, 1988). Preliminary receptor binding studies in renal membrane fractions of medullary cones originating from control and diabetes insipidus chicken (see also Fig. 11.6) also indicate a down-regulation of the medullary AVT receptor from 18 to 8 fmol/mg protein in the presence of high AVT plasma concentrations (McCracken et al., 1994).

REGULATION OF RENAL ADH RECEPTOR

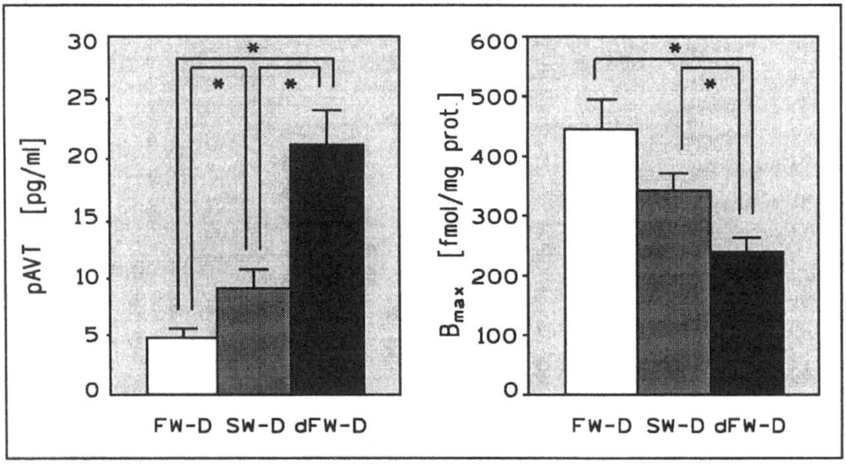

Fig.11.11. AVT plasma concentrations (pAVT) (LEFT PANEL) and corresponding maximum binding capacities (B_{max}) for [^{3}H]AVP to a nephronal membrane fraction (RIGHT PANEL) for freshwater- (FW-D), saltwater- (SW-D) and 24 h dehydrated freshwater-acclimated (dFW-D) ducks. Number of experiments for each group: n=6. Changes in plasma parameters and receptor densities were tested for significance with the unpaired Student's t-test (*: $2P < 0.05$).

For detailed pharmacological characterization of the avian renal ADH receptor system, neurohypophyseal hormones such as AVT, AVP, mesotocin, oxytocin or the hypothalamic fragment [pGlu^1,Cyt^6]AVP$_{(4\text{-}9)}$ were tested for their potency to competitively displace [^{3}H]AVP from its renal binding sites (Fig. 11.10.). With AVT and mesotocin representing avian-specific, and AVP and oxytocin mammalian-specific peptides, the fragment AVP$_{(4\text{-}9)}$ could be detected in both classes of vertebrates (Burbach, 1986). AVP and AVT proved to be equipotent to compete with the radioligand at high affinity, whereas oxytocin and mesotocin revealed potencies 60 and 400 times lower, respectively (Fig. 11.10.). Labelling of general neurohypohyseal peptide receptors in the duck kidney using [^{3}H]AVP as radioligand appear to be ruled out by the very low affinity of mesotocin in the competitive displacement studies. Binding experiments performed for the glomeruli of the anuran *Xenopus laevis* exhibited comparable data for all four neuropeptides (Kloas & Hanke, 1992). Using the tritiated V_2-specific agonist dDAVP as a suitable radioligand to characterize renal V_2-receptors in medullopapillary membranes of the rat kidney, oxytocin proved to have very low affinity for the mammalian V_2-receptor subtype (Marchingo et al., 1988). V_{1a}-vasopressin receptors in the rat cortical collecting duct and cultured mesangial cells, however, showed only 50-100 times lower affinity for oxytocin than AVP to displace non-selective ADH receptor ligands (Ammar et al., 1982; Jard et al., 1987). The AVT/AVP fragment peptide did not bind to putative ADH receptors in the duck nor rat kidney (Marchingo et al., 1988). The avian kidney ADH receptor therefore appears to exhibit ligand binding characteristics quite different from those reported for various mammalian species, and - if anything - revealed a pharmacological profile more closely related to the V_1- than V_2-receptor subtype.

The characterization of binding sites for radiolabelled AVP and AVP analogues revealed the presence of two major AVP receptor subtypes in the rat kidney (Tribollet et al., 1988; Gerstberger & Fahrenholz, 1989). Antidiuretic and antinatriuretic AVP effects in the kidney therefore appear mediated by V_{1a}-(mesangium, cortical thick ascending limb) and V_2-(collecting duct) specific AVP receptors (Ammar et al., 1992; Jard et al., 1987; Leite & Suki, 1990; Rose et al., 1991). Employing numerous AVP receptor subtype-specific antagonists, already tested in the pharmacological experiments in conscious ducks, as well as the V_2-specific agonist dDAVP in duck kidney binding assays (Fig. 11.10.), none of the ligands were able to displace [^{3}H]AVP from its binding sites, with IC50 values at least 2,000 times higher than for AVP or AVT. These data were again comparable to those obtained for the *Xenopus* kidney (Kloas & Hanke, 1992), where V_8- and V_2-receptor subtype-specific ADH analogues (according to their classification in mammals) revealed only low potencies to displace [^{3}H]AVP. Also in the kidney of the brushtail possum, an Australian marsupial, high binding capacity has been observed for tritiated AVP with pharmacological characteristics quite different from the classical V_{1a}- and V_2-receptor subtypes in most mammals (Bathgate and Sernia, 1995). On the other hand, [^{3}H]AVP binding characteristics in the urodele amphibians resembled data obtained for mammalian species (Boyd & Moore, 1990; Fahrenholz et al., 1984; Rose et al., 1991)

The classification of the $V_{1a/b}$- and V_2-receptor subtypes in mammals is based on different ligand specificity as indicated above, but mainly different second messenger formation. Whereas $V_{1a/b}$-subtype-mediated AVP actions are due to phosphoinositol pathway activation with a subsequent rise in intracellular Ca^{2+}-concentration (Jard et al., 1987; Nitschke et al., 1991), C-protein mediated stimulation of adenylate cyclase occurs during V_2-subtype specific interaction (Caltabiano & Kinter, 1991; Rose et al., 1991). Recently, the molecular cloning of the renal V_2-subtype of mammalian AVP receptors has been reported (Birnbaumer et al., 1992; Lolait et al., 1992). All ADH receptors cloned and/or sequenced so far belong to the class of seven transmembrane helix receptors, which lead to cellular responses via G-protein interaction (Mahlmann et al., 1994). To prove the expression of G-protein coupled receptors responsible for AVT-induced renal antidiuresis and -natriuresis, duck kidney membranes were incubated with

AVT-induced renal antidiuresis and -natriuresis, duck kidney membranes were incubated with tritiated AVP in the presence of nonhydrolyzable GTP-analogues, thus not allowing the reassociation of the alpha- and beta/gamma-subunits of dissociated G-proteins, and therefore reducing ligand binding. As demonstrated in Fig. 11.12., binding of $[^3H]$AVP to renal membranes was reduced in the presence of non-hydrolyzable GTP analogues suggestive of G-protein coupled receptor activity

With regard to second messenger formation, incubation of isolated membrane fractions of the duck kidney did not reveal formation of cyclic AMP (cAMP) (Fig. 11.13.). The presence of active adenylate cyclase in this preparation, however, could be demonstrated by a 25-fold elevation of intracellular cAMP after stimulation with forskolin and the dose-dependent responsiveness to parathyroid hormone (data not shown). In contrast, AVT stimulated the formation of inositol trisphosphate (IP3), the second messenger system described for activation of

AVT-INDUCED SIGNAL TRANSDUCTION

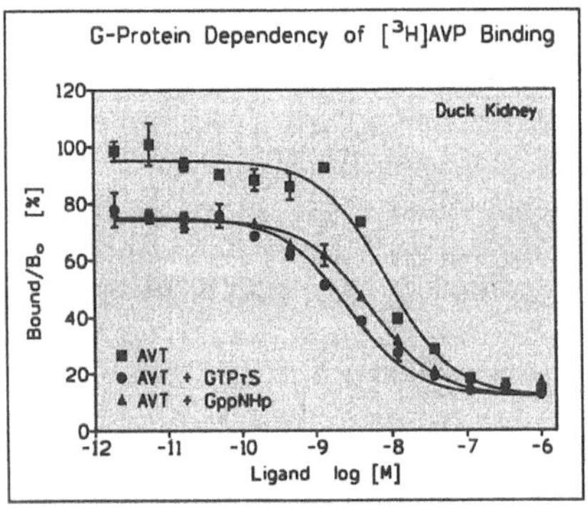

Fig.11.12. G-protein coupling. Displacement of $[^3H]$AVP (2 nM) from binding sites in enriched duck nephronal membrane fraction by increasing concentrations of AVT in the absence (■) or presence of 10^{-4} M non-hydrolyzable GTP-analogues GppNHp (▲) or GTP-gamma-S (●).

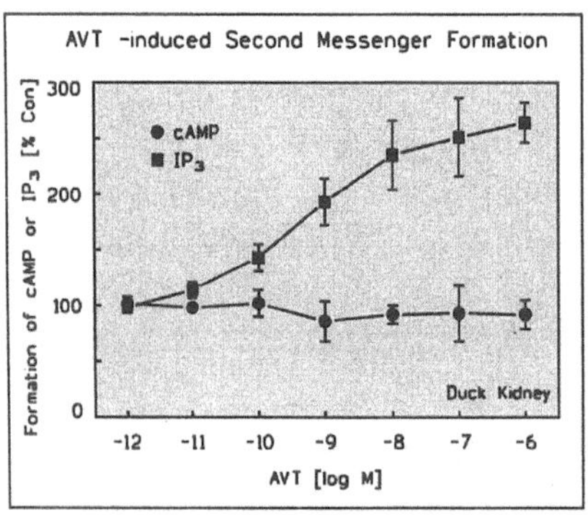

Fig.11.13. Effects of AVT (10^{-12}- 10^{-6} M) on the formation of cyclic AMP (cAMP) and inositol trisphosphate (IP_3) as second messengers using an enriched membrane fraction of duck nephrons.

classical ADH-receptors of the V_1-subtype (Fig. 11.13.). The lack of AVT induced cAMP formation in the duck kidney is supported by findings in the bobwhite quail. Dissected medullary cones of bobwhite quail (*Colinus virginianus*) kidneys incubated with 10-6 M AVT in the absence or presence of phosphodiesterase inhibitors did not reveal responsiveness of the medullary adenylate cyclase system to the neuropeptide, but to application of forskolin (Goldstein, 1995). Quite opposite, and as shown for the collecting duct zone of mammalian kidneys, Stallone and coworkers (1994) could demonstrate, that AVT at the high dose of 10^{-7} M led to a significant increase in intracellular cAMP production in isolated medullary cones of Leghorn fowl. Thus the questions remains open, as to whether avian renal ADH receptors do or do not activate adenylate cyclase. Also cAMP formation by AVT could be demonstrated for kidney preparations of lower vertebrates. In both glomerular and collecting duct regions of the frog kidney (*Rana ridibunda*) obtained from collagenase-treated kidneys, AVT stimulated cAMP formation (Ammar et al., 1995). Various AVT/AVP analogues including dDAVP and V_1- and V_2- -receptor antagonists also tested in the nephronal preparation of the duck kidney did not, however, compete with AVT to enhance adenylate cyclase activity. In nephron segments isolated from both fresh- and seawater-adapted rainbow trout (*Onchorhynchus mykiss*), cAMP led to a dose-dependet rise in intracellular cAMP, and this response was not modified in the presence of atrial natriuretic factor (Perrott et al., 1993). In the same species AVT even at low doses caused a pronounced inhibiton of cAMP in membrane fractions from the gill epithelium (Guibbolini & Lahlou, 1992). This indicates that AVT can produce opposite effects on the second messenger systems in two osmoregulatory important target organs.

11.5 Conclusion

The physiological role of ADH receptors of both the V_{1a}- and V_2-subtype with regard to renal haemodynamics and water elimination in the mammalian kidney became apparent only recently (Rose et al., 1991). Vasopressin V_2 receptors are located along the entire cortical collecting duct, whereas V_1 receptors could be described for cortical collecting duct and glomerular mesangial cells (Burnatowska-Hledin & Spielman, 1989). As demonstrated, receptor binding studies in the duck kidney have not revealed the presence of either of these subtypes employing various V_1- and V_2-pecific receptor antagonists and the V_2-specific agonist dDAVP. Despite the partial antidiuresis but not antinatriuresis caused by dDAVP when administered at high concentrations, (1) the lack of ADH receptor interaction of various V_1- and V_2-specific ligands, (2) the ineffectiveness of V_1- and V_2-specific antagonists to block renal AVT actions and (3) the failure of AVT to stimulate cyclic AMP production make it difficult to decide whether the classical mammalian ADH receptors subtypes (Kirk & Schafer, 1992) exist in the avian kidney or not. Further pharmcological and molecular biological studies might give better insight into the cellular mechanisms of AVT-induced antidiuresis in birds.

References

Amer, S. & Brown, J.A. (1995) Glomerular actions of arginine vasotocin in the *in situ* perfused trout kidney. Am. J. Physiol. 269: R775-R780.
Ammar, A., Roseau, S. & Butlen, D. (1992) Pharmacological characterization of Vla vasopressin receptors in the rat cortical collecting duct. Am. J. Physiol. 262: F546-F553.
Ammar, A., Roseau, S. & Butlen, D. (1995) Vasotocin-sensitive adenylate-cyclase in frog glomeruli. Gen. Comp. Endocrin. 98: 102-110.
Andersson, B. (1978) Regulation of water intake. Physiol. Rev. 58: 582-603.
Bankir, L., Bouby, N. & Trinh-Trang-Tan, M.M. (1989) The role of the kidney in the maintenance of water balance. Bailliere s Clin. Endocr. Metab. 3: 249-311.
Bathgate, R.A.D. & Sernia, C. (1995) Characterization of vasopressin and oxytocin receptors in an Australian marsupial. J. Endocrin. 144: 19-29.
Bie, P. (1980) Osmoreceptors, vasopressin, and control of renal water excretion. Physiol. Rev. 60: 961-1048.
Birnbaumer, M., Seibold, A., Gilbert, S., Ishido, M., Barberis, C. Antaramian, A., Brabet, P. & Rosenthal, W. (1992) Molecular cloning of the receptor for human antidiuretic hormone. Nature 357: 333-335.
Boyd, S.K. & Moore, F.L. (1990) Autoradiographic localization of putative arginine vasotocin receptors in the kidney of a urodele amphibian. Gen. Comp. Endocrin. 78: 344-350.
Braun, E.J. & Dantzler W.H. (1974) Effects of ADH on single-nephron glomerular filtration rates in the avian kidney. Am. J. Physiol. 226: 1-8.
Braun, E.J. (1991) The role of the avian kidney in fluid and electrolyte balance. Acta XXth Congr. Internat. Ornith. 38: 2130-2137.
Brummermann, M. & Braun, E.J. (1995) Renal response of roosters with diabetes insipidus to infusions of arginine vasotocin. Am. J. Physiol. 269: R57-R63.
Burbach, J.P.H. (1986) Proteolytic conversion of oxytocin, vasopressin and related peptides in the brain. In: Neurobiology of Oxytocin (Eds: Ganten, D. & Pfaff, D.), pp. 55-90, Heidelberg, Springer Press.
Burnatowska-Hledin, M.A. & Spielman, W.S. (1989) Vasopressin V_1-receptors on the principal cells of the rabbit cortical collecting tubule. J. Clin. Invest. 83: 84-89.
Caltabiano, S. & Kinter, L.B. (1991) Up-regulation of adenylate cyclase-coupled vasopressin receptors after chronic administration of vasopressin antagonists to rats. J. Pharm. Exp. Ther. 258: 1046-1054.
Cree, A. (1988) Effects of arginine vasotocin on water balance of three leiopelmatid frogs. Gen. Comp. Endocrin. 72: 340-350.
Dantzler, W. (1967) Glomerular and tubular effects of arginine vasotocin in water snakes (*Natrix sipedon*). Am. J. Physiol. 212: 83-91.
Dantzler, W.H. (1980) Comparative nephron function in reptiles, birds, and mammals.Am. J. Physiol. 239: R197-R213.
Dantzler, W.H. (1989) Comparative Physiology of the Vertebrate Kidney. Springer, New York.
Dousa, T.P. (1974) Effects of hormones on cyclic AMP formation in kidneys of nonmammalian vertebrates. Am. J. Physiol. 226: 1193-1197.

Fahrenholz, F., Boer, R., Crause, P., Fritzsch, G. & Grzonka, Z. (1984) Interactions of vasopressin agonists and antagonists with membrane receptors. Eur. J. Pharmacol. 100: 47-58.

Gerstberger, R. & Fahrenholz, F. (1989) Autoradiographic localization of V_1 vasopressin binding sites in rat brain and kidney. Eur. J. Pharmacol. 167: 105-116.

Gerstberger, R. (1996) Hypothalamic control of osmoregulation. In: Perspectives in Avian Endocrinology (Eds: Etches, R. & Harvey, S.), (in press).

Gerstberger, R., Bender, R., Jurzak, M., Keil, R., Kuchenmeister, I. & Schutz, H. (1994) Modulation of avian renal salt and water elimination by arginine vasotocin, mesotocin and ADH receptor subtype-specific analogues. In: Integrative and Cellular Aspects of Autonomic Functions: Temperature and Osmoregulation (Eds: Pleschka, R. & Gerstberger, R.), pp. 533-542, Paris, Eurotext.

Gerstberger, R., Kaul, R., Gray, D.A. & Simon, E. (1985) Arginine vasotocin and glomerular filtration rate in saltwater-acclimated ducks. Am. J. Physiol. 248: F663-F667.

Goldstein, D.L. (1995) Effects of water restriction during growth and adulthood on renal function of bobwhite quail, *Colinus virginianus*. J. Comp. Physiol. B 164: 663-670.

Gray, D.A. & Erasmus, T. (1988) Glomerular filtration changes during vasotocin-induced antidiuresis in kelp gulls. Am. J. Physiol. 255: R936-R939.

Gray, D.A. & Simon, E. (1983) Mammalian and avian antidiuretic hormone: studies related to possible species variation in osmoregulatory systems. J. Comp. Physiol. 151 B: 241-246.

Guibbolini, M.E. & Lahlou, B. (1992) G-I protein mediates adenylate cyclase inhibition by neurohypophyseal hormones in fish gill. Peptides 13: 865-871.

Hamann, D., Hunt, N. & Ivell, R. (1992) The chicken vasotocin gene. J. Neuroendocrin. 4: 505-513.

Jard, S., Lombard, C., Marie, J. & Devilliers, G. (1987) Vasopressin receptors from cultured mesangial cells resemble V_{1a} type. Am. J. Physiol. 253: F41-F49.

Jung, K.Y. & Endou, H. (1991) A novel vasopressin receptor in rat proximal tubule. Biochem. Biophys. Res. Comm. 180: 131 - 137, 1991.

Jurkevich, A., Barth, S. & Grossmann, R. (1996) Sexual dimorphism of Arg-vasotocin expressing neurons in the telencephalon and dorsal diencephalon of the domestic fowl: immunocytochemical and *in situ* hybridization study. Cell Tissue Res.: in press.

Kirk, K.L. & Schafer, J.A. (1992) Water transport and osmoregulation by antidiuretic hormone in terminal nephron segments. In: The Kidney: Physiology and Pathophysiology (Eds: Seldin, D.W. & Giebisch, G.), pp. 1693-1725, New York, Raven Press.

Klempt, M., Ellendorff, F. & Grossmann, R. (1992) Functional maturation of arginine vasotocin secretory responses to osmotic stimulation in the chick embryo and the newborn chicken. J. Endocrinol. 133: 269-274.

Kloas, W. & Hanke, W. (1992) Localization and quantification of nonapeptide binding sites in the kidney of *Xenopus laevis*: Evidence for the existence of two different nonapeptide receptors. Gen. Comp. Endocrin. 85: 71-78.

Leite, M. & Suki, W.N. (1990) AVP and dDAVP in rabbit cortical collecting tubule: a comparative time-course study. Am. J. Physiol. 258: R99-R103.

Lolait, S.J., O'Carroll, A-M., McBride, O.W., Konig, M., Morel, A. & Brownstein, M. (1992) Cloning and characterization of a vasopressin V_2 receptor and possible link to nephrogenic diabetes insipidus. Nature 257: 336-339.

Mahlmann, S., Meyerhoff, W., Hausmann, H., Heierhorst, J., Schonrock, C., Zwier, H., Lederis, K. & Richter, D. (1994) Structure, function, and phylogeny of [Arg8]vasotocin receptors from teleost fish to toad. Proc. Natl. Acad. Sci. USA 91: 1342-1345.

Marchingo, A.J., Abrahams, J.M., Woodcock, E.A., Smith, A.I., Mendelsohn, F.A.O. & Johnston, C.I. (1988) Properties of (^{3}H)-1-Desamino-8-D-arginine vasopressin as a radioligand for vasopressin V_2-receptors in rat kidney. Endocrinology 122: 1328-1336.

McCracken, D., Dantzler, W.H. & Braun, E.J. (1994) Binding of antidiuretic hormone in renal cell membrane fragments from normal and nephrogenic diabetes insipidus (NDI) domestic fowl. FASEB J. 8: 4838 (Abstract).

Morild, I., Bohle, A. & Christensen, J.A. (1985) Structure of the avian kidney. Anatom. Rec. 212: 33-40.

Muhlbauer, E., Hamann, D., Xu, B., Ivell, R., Ellendorff, F. & Grossmann, R. (1992) Arginine vasotocin gene expression during osmotic challenge in the chicken.J. Neuroendocrin. 4: 347-351.

Nishimura, H., Miwa, T. & Bailey, J.R. (1984) Renal handling of sodium chloride and its control in birds. J. Exp. Zool. 232: 697-705.

Nitschke, R., Frobe, U. & Greger, R. (1991): Antidiuretic hormone acts via V_1 receptors on intracellular calcium in the isolated perfused rabbit cortical thick ascending limb. Eur. J. Physiol. 417: 622-632.

Pang, P.K.T., Uchiyama, M. & Sawyer, W.H. (1982) Endocrine and neural control of amphibian renal functions. Fed. Proc. 41: 2365-2373.

Perrott, M.N., Sainsbury, R.J. & Balment, R.J. (1993) Peptide hormone-stimulated messenger production in the Teleostean nephron. Gen. Comp. Endocrin. 89: 387-395.

Rose, C.E., Rose, K.Y. & Kinter, L.B. (1991) Effect of V_1/V_2-receptor antagonism on renal function and response to vasopressin in conscious dogs. Am. J. Physiol. 260: F273-F282.

Schutz, H. & Gerstberger, R. (1992) Modulation of kidney function in conscious Pekin ducks by atrial natriuretic factor. Endocrinology 130: 678-684).

Simon, E. (1982) The osmoregulatory system of birds with salt glands. Comp. Biochem. Physiol. 71A: 547-556.

Simon, E., Gerstberger, R. & Gray, D.A. (1992) Central angiotensin II responsiveness in birds. Prog. Neurobiol. 39: 179-207.

Stallone, J. & Braun, E.J. (1985) Contributions of glomerular and tubular mechanisms to antidiuresis in conscious domestic fowl. Am. J. Physiol. 249: F842-F850.

Stallone, J.N. & Braun, E.J. (1986) Osmotic and volume regulation of plasma arginine vasotocin in conscious domestic fowl. Am. J. Physiol. 250: R644-R657.

Stallone, J.N., Pacelli, M.M. & Braun, E.J. (1994) Arginine vasotocin (AVT) and forskolin (FSK) stimulate medullary cyclic AMP (cAMP) in the avian kidney. FASEB J. 8: 4387 (Abstract).

Steiner, M. & Phillips, M.I. (1988) Renal tubular vasopressin receptors downregulated by dehydration. Am. J. Physiol. 254: C404-C410.

Tribollet, E., Barberis, C., Dreifuss, J.J. & Jard, S. (1988) Autoradiographic localization of vasopressin and oxytocin binding sites in the rat kidney. Kidney Int. 33: 959-965.

Verbalis, J.G., Mangione, M.P. & Stricker, E.M. (1991) Oxytocin produces natriuresis in rats at physiological plasma concentrations. Endocrinology 128, 1317-1322.

Wideman, R.F. (1988) Avian kidney anatomy and physiology. CRC Crit. Rev. Poultry Biol. 1: 133-176.

Wilson, X.J. (1984) The renin-angiotensin system in nonmammalian vertebrates. Endocr. Rev. 5: 454-61.

Intracellular signalling in salt-secreting cells - recent advances in the avian nasal gland model

T.J. Shuttleworth, J.L. Thompson and S.C. Martin

Department of Pharmacology and Physiology, University of Rochester School Medicine and Dentistry, Rochester, NY 14642, USA

12.1 Introduction

The avian nasal gland has had an important role as a model system in studies of ion secretion for almost forty years, ever since the discovery of its role as an extra-renal salt-secreting organ by Schmidt-Nielsen and colleagues (Fänge et al. 1958). Less widely appreciated is the fact that this tissue also played a critical role in the earliest studies of the intracellular signalling pathways involved in the stimulation of cell activity. Some of the original studies demonstrating for the first time that the muscarinic stimulation of secretory activity in cells was associated with an increased turnover of membrane phosphoinositides were performed on the avian nasal gland (Hokin and Hokin 1964). Unfortunately, the key connection between this turnover of phosphoinositides and increases in cytosolic Ca^{2+} concentrations ($[Ca^{2+}]_i$) was not appreciated until more than ten years later (Michell 1975), and the role of inositol 1,4,5-trisphosphate in providing the critical link between these two processes was not revealed for a further eight years (Streb et al. 1983).

The basic mechanism of secretion seen in the nasal gland is a very common one involving the secondary active transepithelial transport of chloride (see below). It is largely for this reason that this tissue has been extensively used as a model system for the epithelial secretion of ions and fluid in general (e.g. in exocrine glands). An additional advantage of the avian nasal gland as a model tissue is that its only function is the secretion of ions and fluid. Analysis of the secretory mechanism and its control is not therefore complicated by the secretion of other components such as proteins or mucus that typifies the activity of most of the other exocrine tissues. The avian nasal gland is therefore an extremely homogeneous tissue that can readily be dissociated into single cells for the detailed analysis of the cellular mechanisms involved in the secretory process and their control. The successful application of such powerful techniques as patch-clamp electrophysiology, and single cell digital microfluorimetric analysis of cytosolic Ca^{2+} and pH to these cells, along with biochemical assays of intracellular second messenger generation and turnover, and secretory activity, has meant that the avian nasal gland has become a versatile and valuable system for the study of the mechanisms of ion secretion and their control.

12.2 Mechanism of Secretion

As noted above, the underlying mechanism of ion and fluid secretion is consistent with that seen in a variety of different tissues such as various exocrine cells, as well as certain epithelia of the respiratory tree and of the intestine for example (Frizzell et al. 1979). The overall transport mechanism (Fig.12.1) is driven by a basolateral Na-K-pump (Na-K-ATPase) that generates a low-sodium environment in the cell providing

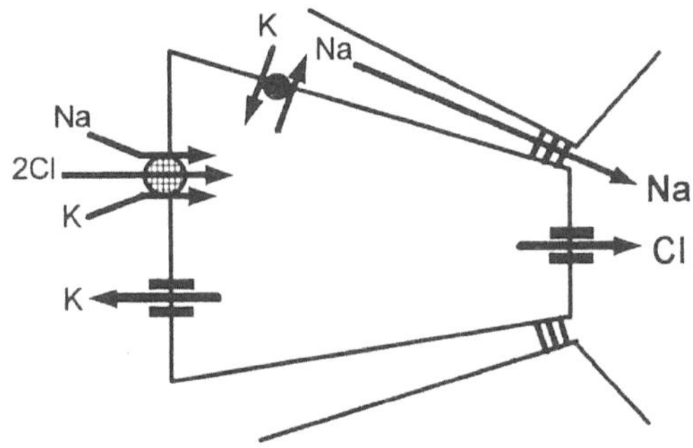

Fig. 12.1. Diagram illustrating the basic model of ion and fluid secretion by the cells of the avian nasal gland. See text for details.

a driving force for the entry of sodium from the basolateral side. This entry of sodium is principally coupled to the uphill entry of chloride via a Na-2Cl-K cotransporter that is sensitive to loop-diuretics, such as bumetanide, thereby effectively accumulating chloride in the cell above its electrochemical equilibrium, creating the conditions whereby chloride can exit the cell to the luminal side via apical chloride channels. Potassium ions, in the meantime, simply recycle across the basolateral membrane via appropriate channels. This transcellular movement of chloride ions results in the development of a lumen-negative transepithelial potential which acts to pull sodium ions via a paracellular path in the basal to luminal (apical) direction. The resulting secretion of sodium and chloride generates the osmotic force for the associated secretion of fluid. The evidence for the basic applicability of this model to the avian nasal gland is extensive and will not be repeated here (see Shuttleworth 1995 for references).

Largely based on information obtained from a variety of chloride-secreting tissues, particularly several of the more "classic" exocrine glands (pancreas, salivary, lacrimal), it is generally considered that the primary sites of activation of this transport mechanism are the membrane ion channels (i.e. the apical Cl^- conductance and/or the basolateral K^+ conductance) (see Petersen 1992), but it is only fairly recently that this has been examined in the avian nasal gland (see below). It is also important to note that more long-term activation of secretion probably involves other components of the transport mechanism such as the basolateral Na-2Cl-K cotransporter and the Na-K-ATPase, and evidence for just such effects has been reported (Torchia et al. 1991). However, it is clear that the immediate responses to neurotransmitter and/hormone action are essentailly limited to the activation of the ion channels, and it is this aspect that forms the focus of this review.

12.3 Intracellular Calcium Signalling

The principal regulator of secretory activity is acetylcholine released from the parasympathetic nerve endings of the so-called "secretory nerve" (Peaker and Linzell 1975). Acetylcholine activates muscarinic receptors on the secretory cell that are coupled to the enzyme phospholipase C via a member of the G_q family of G proteins (Hildebrandt and Shuttleworth 1993). The activation of phospholipase C results in the breakdown of the membrane phospholipid phosphotidylinositol 4,5-bisphosphate to liberate two second messengers, diacylglycerol which is hydrophobic and remains in the membrane domain, and the water-soluble inositol 1,4,5-trisphosphate ($InsP_3$) which enters the cytosol (Shuttleworth and Thompson 1989). The generated $InsP_3$ binds onto specific $InsP_3$-receptors located in the membranes surrounding certain intracellular Ca^{2+} stores that apparently comprise part of the endoplasmic reticulum (Mikoshiba 1993; Shuttleworth 1994). The effect of $InsP_3$ binding onto its receptor is to open a cation channel, which is an integral part of the receptor molecule (Bezprozvanny and Ehrlich 1995), resulting in the release of Ca^{2+} from the store into the cytosol, thereby initiating the "Ca^{2+} signal". Because of the finite capacity of the intracellular Ca^{2+} stores mobilized by $InsP_3$, this sequence of events results in only a transient "spike" of elevated cytosolic Ca^{2+} concentrations. The sustained elevation in $[Ca^{2+}]_i$ typical of cells undergoing maximal stimulation is the result of the entry of Ca^{2+} from the extracellular medium which, it is generally assumed, occurs via some kind of Ca^{2+} channel. The well-characterized family of voltage-activated Ca^{2+} channels are usually absent in non-excitable cells, so the specific nature of the relevant channel(s) and exactly how they are activated is unclear. Currently, models for the receptor-induced activation of this Ca^{2+} entry in non-excitable cells invoke the so-called "capacitative" mechanism in which it is the emptying of the agonist-sensitive stores *per se* that induces the activation of the Ca^{2+} entry pathway (Putney 1986; Putney 1990). Evidence supporting the presence of such a mechanism is extensive for a wide variety of non-excitable cells, and recently several studies have identified a Ca^{2+} conductance in the plasma membrane that is specifically activated by depletion of the stores (Hoth and Penner 1992, 1993; Zweifach and Lewis 1995a, 1995b). However, exactly how information regarding the Ca^{2+} status of the intracellular stores is relayed to the plasma membrane channels remains unknown.

In recent years, following the development of systems for studying the Ca^{2+} responses in single individual cells, it has become apparent that the agonist-induced $[Ca^{2+}]_i$ signals in many cells are far more complex than was originally supposed. Rather than the simple elevation to a new sustained $[Ca^{2+}]_i$ value ("plateau response") commonly seen at supramaximal agonist concentrations, physiological levels of agonists typically induce a series of marked oscillations in

$[Ca^{2+}]_i$ which often continue at essentially constant frequency and amplitude as long as agonist is present (Fig. 12.2). It is clear that these oscillations can arise by a variety of different mechanisms in different cells (Berridge, 1990) but,

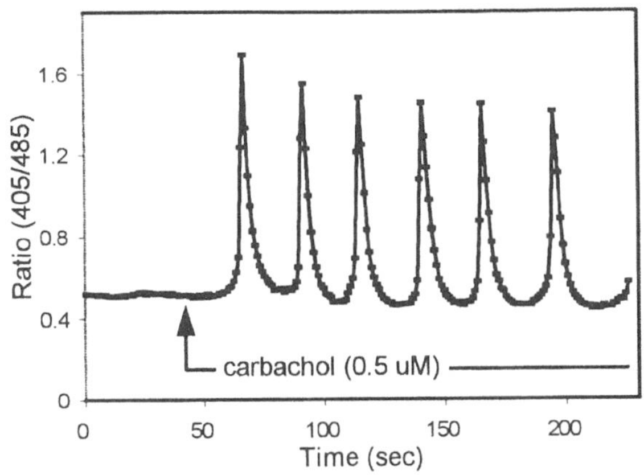

Fig. 12.2. Oscillations in intracellular Ca^{2+} concentration in a single avian nasal gland cell in response to a low concentration (0.5 mM) of the muscarinic agonist carbachol. Intracellular Ca^{2+} concentration was recorded by photon-counting microfluorimetry of intracellularly loaded indo-1 and is presented as the emission ratio at 405nm and 485nm (excitation at 350nm). Data redrawn from Shuttleworth and Thompson (1996a).

in many cases, they reflect the unique biphasic behavior of the $InsP_3$-receptor to changes in $[Ca^{2+}]_i$ (De Young and Keizer 1992). At low concentrations, cytosolic Ca^{2+} acts as a co-agonist with $InsP_3$ on its receptor resulting in a positive feedback between $InsP_3$ itself and the released Ca^{2+} (Iino 1990; Finch et al. 1991; Bezprozvanny et al. 1991). At higher concentrations (generally >300 nM), a more slowly developing Ca^{2+}-dependent inactivation or desensitization is revealed which acts to close the $InsP_3$ channel. Together, these actions result in the repetitive release and re-uptake of Ca^{2+} from the $InsP_3$-sensitive stores. This mechanism certainly explains the $[Ca^{2+}]_i$ oscillations induced in avian nasal gland cells by low concentrations of the muscarinic agonist carbachol. In these cells the evidence indicates that other Ca^{2+} stores, such as those sensitive to caffeine and ryanodine and which are believed to be involved in $[Ca^{2+}]_i$ oscillations in certain other cell types, are absent (Martin and Shuttleworth 1994a). No matter what the precise details of the oscillatory mechanism, in essentially all cases the underlying mechanism is inherently intrinsic to the cell with the consequence that possible roles of Ca^{2+} entry in such oscillatory responses have largely been ignored. Nevertheless, $[Ca^{2+}]_i$ oscillations are usually dependent, to at least some degree, on the presence of an extracellular source of Ca^{2+} and changes in the rate of Ca^{2+} entry have been shown to have rather immediate effects on oscillation frequency in several cells including those from the avian nasal gland (Shuttleworth and

Thompson 1996a). Until recently, such effects were generally attributed to a requirement for Ca^{2+} entry to recharge the stores sufficiently during each interspike interval before release could occur (Berridge 1993, 1994; Thorn 1995). However, recent evidence has shown that this is unlikely to be the case. In the avian nasal gland, we have shown that restricting the ability of the stores to refill (simply by modestly reducing extracellular Ca^{2+}) did not prevent their repetitive emptying in an oscillatory manner and the oscillations continued with a declining amplitude until the stores eventually emptied completely (Shuttleworth and Thompson 1996a). Clearly, the ability of the stores to release Ca^{2+} in an oscillatory manner is unrelated to their filling status. Furthermore, we could show that under normal circumstances the stores are essentially fully recharged as soon as $[Ca^{2+}]_i$ has returned to baseline immediately after each oscillation. This demonstrates that the interspike interval does not reflect a period during which the stores are recharging. From these data, we concluded that the role of Ca^{2+} entry in modulating oscillation frequency was unlikely to be related to any role in refilling the stores between spikes (Shuttleworth and Thompson 1996a). The alternative explanation was that Ca^{2+} entry was influencing oscillation frequency by affecting the release of Ca^{2+} from the stores and, consistent with this, we found that acute inhibition of Ca^{2+} entry by a variety of different means resulted in the immediate cessation of the oscillations (Martin and Shuttleworth 1994b). Significantly, the oscillations stopped with the stores full, not empty. These data showed that Ca^{2+} entry was indeed involved in inducing the release of Ca^{2+} from the $InsP_3$-sensitive stores. They further indicated that, at the low concentrations of agonist that give rise to oscillatory $[Ca^{2+}]_i$ responses, the generated levels of $InsP_3$ are *in themselves* inadequate to produce the release of Ca^{2+} from the agonist-sensitive stores. The likely explanation for these data is that the critical role that Ca^{2+} entry plays in the oscillatory responses involves it acting as a co-agonist with $InsP_3$ on the $InsP_3$ receptor, thereby triggering the repetitive release of the Ca^{2+} from the stores (Shuttleworth and Thompson 1966a; Taylor and Traynor 1995). Interestingly, we have recently obtained evidence indicating that the nature of this Ca^{2+} entry during oscillations demonstrates characteristics that are incompatible with the capacitative model described above, suggesting that it involves a distinct agonist-induced activation mechanism perhaps more directly linked to the muscarinic receptor (Shuttleworth and Thompson 1996b).

12.4 Activation of Secretion by Intracellular Calcium

To investigate the relevant targets for these oscillatory changes in $[Ca^{2+}]_i$ in the secretory mechanism, we have utilized the perforated-patch configuration of the whole-cell patch clamp technique. In this, a patch pipette that has been back-filled with the pore-forming antibiotic amphotericin B is applied to the cell surface to form a high resistance "giga-seal" in the normal manner. Electrical access to the cytosol is then achieved by the diffusion and insertion of the amphotericin molecules into the membrane patch, resulting in a decline in access resistance, the progress of which can be readily monitored by the increase in the membrane capacitance. The amphotericin pores within the patch provide control over the monovalent ion concentrations in the cytosol, but the normal Ca^{2+}-buffering systems of the intracellular environment are maintained and the wash-out of other large molecules is avoided. Specifically, we found that this technique was critical for the demonstration of the external Ca^{2+} dependency of $[Ca^{2+}]_i$-induced currents during oscillations, and in avoiding the frequently reported run-down of Ca^{2+}-activated Cl^- currents. Using this technique, we found that application of low concentrations of the muscarinic agonist carbachol resulted in rapid transient oscillatory increases in whole-cell membrane currents (Martin and Shuttleworth 1994a). These currents were in the inward direction at potentials more negative than approximately -40 mV, and outward at potentials more positive than -40 mV. Similar overall conductance changes could be illicited by application of the Ca^{2+} ionophore A23187 except that, in this case of course, the currents were sustained not oscillatory. Furthermore, the $[Ca^{2+}]_i$-dependency of the carbachol-induced currents was confirmed by loading the cells with the Ca^{2+}-chelator BAPTA which completely abolished the carbachol-induced currents (Martin and Shuttleworth 1994a). The sustained nature of the Ca^{2+}-dependent currents seen in the presence of A23187 permitted analysis of their composition which would have been difficult with the rapid oscillatory responses seen with carbachol. Application of the K^+ channel blocker TEA markedly reduced the outward Ca^{2+}- activated currents seen at potentials above -40 mV, but was essentially without effect on the inward currents at more negative potentials (Martin and Shuttleworth 1994a). Subsequent addition of the Cl^--channel blockers DPC or NPPB reduced both the inward currents and the residual outward currents to essentially control values. These data indicate that, in the presence of A23187, two separate currents are activated, a TEA-sensitive K^+ current and a DPC/NPPB-sensitive Cl^- current. This can be clearly demonstrated by examination of the currents recorded at 0 mV (Fig. 12.3). Under the conditions of these experiments, the equilibrium potential for potassium (E_K) is approximately -88 mV whilst that for chloride (E_{Cl}) is -31 mV, and therefore currents measured at 0 mV would be expected to comprise a mixture of both K^+ and Cl^- currents. Whole-cell currents measured at this potential are markedly increased by both carbachol and A23187. Addition of TEA reduces the A23187-stiumulated currents by some 60-70%, whilst subsequent addition of DPC reduces the currents to resting levels. The conclusion is therefore that the oscillatory increases in $[Ca^{2+}]_i$ induced by physiological activation of the muscarinic receptors on the secretory cells results in a corresponding activation of both chloride channels and potassium channels. The former are sensitive to inhibition by NPPB and DPC, whilst the latter are inhibited by TEA. Subsequent studies showed that similar currents were activated by other agonists whose actions also resulted in increases in $[Ca^{2+}]_i$. For example, ATP is commonly co-released with acetylcholine from parasympathetic nerve terminals, and it was found that this acted on nucleotide receptors of the P_{2U} class on the secretory cells to increase $[Ca^{2+}]_i$ and activate both $K^+$and Cl^- currents (Martin and Shuttleworth 1995). Recently we have further characterized the Ca^{2+}-activated K^+ channels from

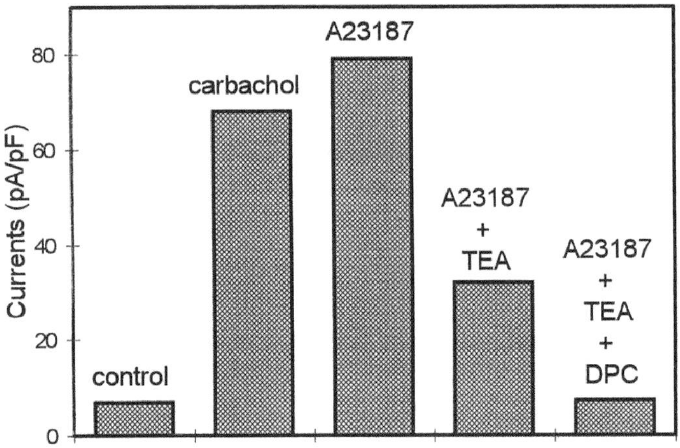

Fig. 12.3. Whole-cell current measurements (in pA/pF) of avian nasal gland cells measured at a holding potential of 0 mV using the perforated patch technique. Effects of increasing intracellular Ca^{2+} by either carbachol or the Ca^{2+} ionophore A23187. At this holding potential the currents measured are predicted to comprise a mixture of K^+ currents (E_K= -88 mV) and Cl^- currents (E_{Cl}= -31 mV). This is confirmed by their inhibition by a combination of the K^+-channel blocker TEA and the Cl^--channel blocker DPC.

avian nasal gland cells by isolating them in preparations of the basolateral membrane and reconstituting them into planer lipid bilayers where their single channel properties can be studied in isolation (Wu et al. 1996). This study (done in collaboration with the laboratory of Dr. Per Stampe) revealed that, as expected, these channels were a member of the so-called "maxi-K^+" family of large conductance Ca^{2+}-activated K^+ channels. As such, they were sensitive to inhibition by TEA and were acutely sensitive to charybdotoxin. Overall, the Ca^{2+}-dependency of the channels showed wide variation but closer examination of numerous records revealed that the individual sensitivity to Ca^{2+} fell into five clusters. The maxi-K^+ channel is known to be a tetrameric protein and it was hypothesized that this clustering of Ca^{2+} sensitivities may reflect the formation of each tetrameric channel from a random mixing of two distinct subunits possessing high and low Ca^{2+} sensitivity respectively (Wu et al. 1996).

Although in isolated cell preparations it is impossible to determine where in the membrane these Cl^- and K^+ channels are located, for them to effect a sustained increase in secretory activity it is reasonable to assume that the Cl^- channels are in the apical membrane whilst the K^+ channels are in the basolateral membrane.

Activation of apical Cl^- channels would permit chloride ions, accumulated in the cytosol as a result of the action of the basolateral cotransporter, to exit to the lumen. At the same time, the opening of the basolateral K^+ channels would keep the cell membrane potential more negative than the chloride equilibrium potential and hence maintain the driving force for the apical chloride exit. The net loss of both K^+ and Cl^- from the cytosol will result in a decrease in cell volume as water follows, and evidence from other systems indicates that this and/or the fall in cytosolic Cl^- concentration is the immediate signal for the activation of the basolateral cotransporter thereby bringing in more K^+ and Cl^- (Lytle and Forbush 1996).

12.5 Activation of Secretion by Intracellular Cyclic AMP

The $[Ca^{2+}]_i$-mediated pathway for the activation of secretion described above is essentially consistent with the mechanisms reported from a variety of other exocrine cell types (Petersen 1992). However, it has long been known that the secretory cells of the avian nasal gland can be stimulated by other agonists whose receptors are coupled to different intracellular signalling pathways, most particularly to the generation of cyclic AMP (Lowy et al. 1985; Shuttleworth and Thompson 1987). Such agonists include VIP, and the β-adrenergic receptor agonist isoprenaline (Lowy et al. 1987; Lowy and Ernst 1987; Gerstberger and Gray 1993; Martin and Shuttleworth 1994c; Martin et al. 1994). To investigate the cellular actions of such agonists we again employed the perforated patch-clamp technique as described above (Martin and Shuttleworth 1994c; Martin et al. 1994). In cells held at -60 mV and alternately pulsed to - 88 mV (the K^+ equilibrium potential), -31 mV (the Cl^- equilibrium potential), 0 mV, and +30 mV, addition of VIP induced a gradual increase in membrane conductance which was preceded by a significant lag of up to 2 minutes (Martin and Shuttleworth 1994c). The VIP-stimulated currents were sustained, showing no indication of the oscillations typical of the responses elicited by carbachol or by ATP, and were in the inward direction at -88 mV and -60 mV, and outwards at 0 mV, +30 mV. Only very small currents were recorded at -31 mV under stimulated conditions. Essentially identical data were obtained with isoprenaline, the effects of which was completely blocked by the receptor antagonist propranolol (Martin et al. 1994). Pre-loading the cells with the Ca^{2+}-chelator BAPTA failed to have any effect on these responses indicating that the isoprenaline-induced responses did not involve changes in $[Ca^{2+}]_i$, a fact confirmed by microfluorimetric determinations of $[Ca^{2+}]_i$ in indo-loaded cells. Similar current responses were observed on application of the adenylyl cyclase activator forskolin, or the exogenous membrane-permeant cAMP anolog 8-cpt-cAMP. Furthermore, the VIP-induced current responses were irreversibly inhibited by the inhibitory cAMP isomer Rp-cAMP (Martin and Shuttleworth 1994c). As to the underlying nature of these responses, examination of the currents measured at the equilibrium potential for K^+ (-88 mV) showed similar increases were induced by carbachol, isoprenaline, and VIP (Fig. 12.4A).

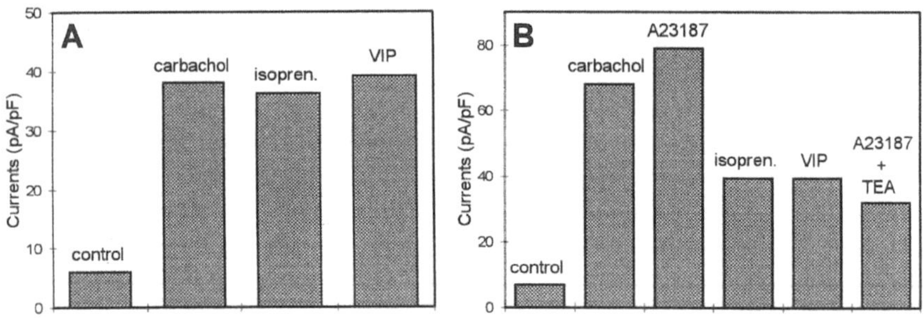

Fig. 12.4. **(A)** Whole-cell current measurements (in pA/pF) of avian nasal gland cells measured at a holding potential of -88 mV. Comparison of the effects of the Ca^{2+}-mobilizing agonist carbachol and the cAMP-generating agonists isoprenaline and VIP. At this potential (= E_K), measured currents are predicted to comprise mainly of Cl^- current **(B)** Whole-cell current measurements (in pA/pF) measured at a holding potential of 0 mV. Comparison of agents that increase intracellular Ca^{2+} with those that increase cAMP levels.

Such currents measured at the equilibrium potential for K^+ most likely reflect the activity of Cl^- channels and the data indicate a similiar stimulation of such channels can be achieved by elevations in $[Ca^{2+}]_i$ and by increases in cAMP. Currents measured at 0 mV which, as noted above, will comprise both K^+ currents and Cl^- currents, showed a rather different picture however (Fig. 12.4B). Here the currents activated by elevations in $[Ca^{2+}]_i$ were far greater than those seen following application of maximal concentrations of agonists that raise intracellular cAMP. However, addition of TEA to block the Ca^{2+}-activated K^+ channels reduced the total Ca^{2+}-activated current to a level comparable to that seen with cAMP. In contrast the cAMP-activated currents were completely unaffected by TEA. From these data we concluded that intracellular cAMP acts specifically to increase Cl^- currents in a sustained manner and, in marked contrast to the effects of increases in $[Ca^{2+}]_i$, is essentially without obvious effect on K^+ channels (Martin et al. 1994). Consistent with this interpretation, changes in the internal (cytosolic) chloride concentration produced shifts in the reversal potential for both VIP-induced and isoprenaline-induced currents that were consistent with those predicted by the Nernst equation assuming that the membrane was behaving as a simple chloride electrode.

As pointed out above, the basolateral K^+ conductance would seem to have a critical role in the overall transport mechanism in that it effectively hyperpolarizes the cell membrane during stimulation thereby preventing the cell from simply going to the equlibrium potential for chloride when the apical Cl^- channels open. Only in this way will the driving force for the luminal exit of chloride ions be maintained. On this basis, the apparent failure of the cAMP-generating agonists to activate any obvious K^+ conductance might be expected to severely impact their ability to effectively stimulate secretory activity by the cells. Indeed, measurements of cell membrane potential reveal that during stimulation by isoprenaline this falls from resting values of around -50 mV to a value close to the equilibrium potential for chloride (-31 mV) (Martin et al. 1994). This clearly raises questions as to how effective these agonists that act via increases in cAMP may be at actually stimulating secretion. As will be discussed below, the principal role of such agonists in the overall control of secretory activity may not be to stimulate secretion *per se*, but rather to markedly potentiate the actions of other agonists acting via $[Ca^{2+}]_i$-mediated signals.

12.6 Interactions between Calcium and Cyclic AMP Signals

To investigate and compare the actual ability of these signals to stimulate secretion in a system that could be correlated to the conditions under which we had made our determinations of changes in $[Ca^{2+}]_i$, in cAMP and the activation of the various membrane currents, it was necessary to use an indirect measure of secretory activity. As can be seen from the basic model described above, the overall transport mechanism is driven by the activity of the sodium pump. The activity of this pump can be assessed in dispersed cells by determining the proportion of the total oxygen consumption that is sensitive to the Na-pump inhibitor ouabain. The problem with this as a measure of transport (or secretory) activity is that the Na-pump activity can be turned on by any process that increases sodium entry into the cell, whether or not this sodium entry is involved in, or related to, the actual secretory mechanism (Shuttleworth 1982). A better measure of secretory activity is to determine the bumetanide-sensitive oxygen consumption. This accurately reflects that component of the overall cell metabolism that is dependent on the basolateral cotransporter, the activity of which is directly associated with the secretory mechanism *per se*. Using this approach, we were able to show that the muscarinic agonist carbachol stimulates secretory activity in a concentration-dependent fashion over the range 1-500 mM, with maximal stimulation (achieved at carbachol concentrations of 500 mM) representing a greater than 20-fold increase in bumetanide-sensitive oxygen consumption over control levels (Martin et al. 1994) (Fig. 12.5). The data also show that isoprenaline is indeed able to stimulate secretory activity, but maximal concentrations of this agonist only result in a less than 6-fold increase in bumetanide-sensitive oxygen consumption. Importantly, in the presence of low concentrations of carbachol (as little as 1 mM), isoprenaline is able to stimulate secretory activity to virtually the same levels as a maximal concentration of carbachol (Martin et al. 1994). These data raise two immediate questions: first, how is an agonist like isoprenaline that acts via increasing cAMP able to stimulate secretory activity, given that cAMP does not appear to activate any K^+ conductance which, as discussed above, would appear to be critical for the overall transport mechanism? Second, what is the basis of the obvious profound potentiation of secretory activity seen with combinations of carbachol and isoprenaline?

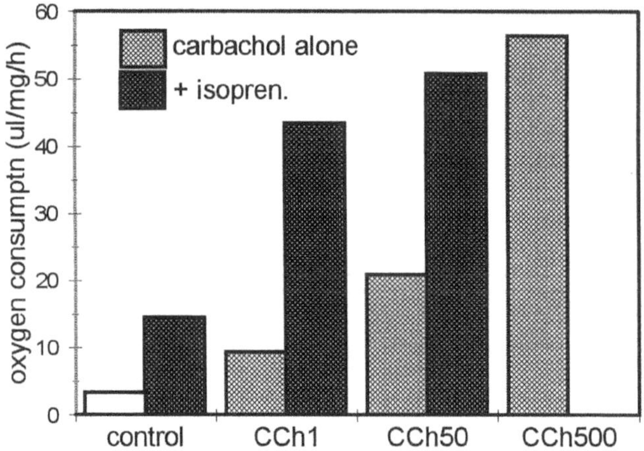

Fig. 12.5. Bumetanide-sensitive oxygen consumption in isolated cells of the avian nasal gland. Measurements were made on suspensions of cells at different concentrations of carbachol either alone or in the presence of a maximal concentration of isoprenaline (10 mM).

That isoprenaline does indeed stimulate secretory activity is supported by previous reports of short-circuit current measurements made on cultured monolayers of nasal gland cells (Lowy and Ernst 1987). Clearly, for such secretory activity, the cell must be able to maintain a small but significant driving force for Cl^- efflux in the presence of isoprenaline. This is obviously not derived from the TEA-sensitive "maxi-K^+" channel activated by Ca^{2+}-mobilizing agonists as the isoprenaline-stimulated, cAMP-dependent currents were unaffected by TEA. Similarly, TEA was without effect on the isoprenaline-induced stimulation of bumetanide-sensitive oxygen consumption. Nevertheless, potassium is effectively the only ion whose equilibrium potential is significantly more negative than chloride, and the cotransporter is electroneutral, so this suggests that a K^+ conductance of some kind must be involved. The most likely candidate appears to be an inwardly rectifying K^+ conductance previously identified in cell-attached recordings in avian nasal gland cells under resting conditions (Richards et al. 1989). This conductance is not activated by Ca^{2+}, and is insensitive to TEA but it is, however, blocked by barium. Consistent with a role for this conductance in the isoprenaline-induced increase in secretory activity, addition of barium completely blocked the isoprenaline-stimulated bumetanide-sensitive oxygen consumption. Presumably the activity of this channel, although not increased during cAMP-dependent stimulation, is sufficient to maintain a small driving force for the exit of chloride following opening of the apical Cl^- channels in the presence of isoprenaline.

As for the potentiation effects observed, we have already described how agonist-induced increases in either $[Ca^{2+}]_i$ or cAMP were both able to activate a Cl^- conductance. However, it was unclear whether these two different cell signals were activating the same, or different, Cl^- channels. Examination of the kinetic properties of the Cl^- conductances activated by cAMP and Ca^{2+} to a range of voltage pulses (from -100 mV to +100 mV) revealed clear differences in their behavior (Martin et al. 1994) (Fig. 12.6). As has been reported for a variety

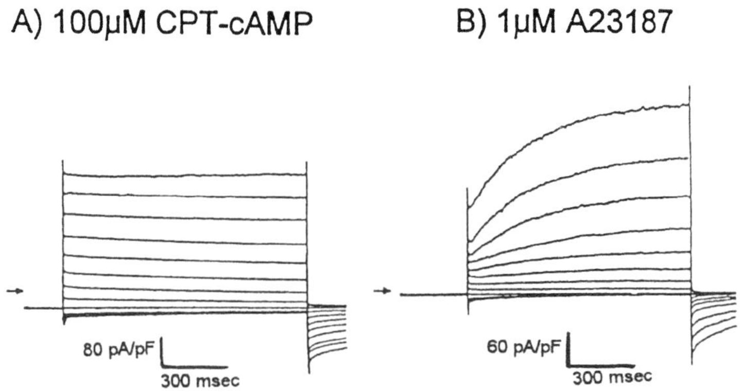

Fig. 12.6. Comparison of the kinetics and current-voltage relationships of whole-cell Cl^- currents in avian nasal gland cells activated by increases in intracellular cAMP (by the permeant metabolically resistant anolog 8-chlorophenylthio-cAMP) and by increases in intracellular Ca^{2+} (via the ionophore A23187). Cells were treated with TEA to block K^+ currents and then pulsed to voltages between -100 mV to +100 mV in 20 mV steps from a holding potential of -60 mV. (Martin et al. 1994)

of other tissues,the Ca^{2+}-dependent Cl^- current shows a marked time-dependent activation and a non-linear current-voltage relationship. In contrast, the cAMP-dependent current shows no time-dependent activation and a far more linear current-voltage relationship. Similar data were obtained with isoprenaline as well as with VIP (Martin and Shuttleworth 1994c). This distinction is confirmed by studies of their respective pharmacology. As noted above, the Ca^{2+}-activated chloride current was inhibited by DPC and by NPPB. However, these agents were without effect on the cAMP-activated chloride current which was however inhibited by flufenamic acid and by niflumic acid. The clear conclusion is that Ca^{2+} and cAMP are activating two distinct Cl^- channels. Consequently, combinations of a $[Ca^{2+}]_i$-mobilizing agonist and a cAMP-generating agonist would be expected to increase the overall Cl^- current significantly more than either agonist alone. Furthermore, in the presence of a $[Ca^{2+}]_i$-mobilizing agonist, the activation of the TEA-sensitive Ca^{2+}-dependent "maxi-K^+" channels, would provide a significant driving force for chloride exit via *both* types of Cl^- channel as well as increasing K^+ exit and hence increasing the turnover of the basolateral cotransporter (Fig. 12.7). All these multiple interacting effects would likely be further increased by the sustained nature of the cAMP-activated Cl^- conductance, compared to the oscillatory behavior of the Ca^{2+}-activated Cl^- conductance (reflecting the oscillatory nature of the agonist-induced changes in $[Ca^{2+}]_i$). An important point to note is that the marked potentiation of secretory activity observed with combinations of Ca^{2+}-mobilizing and cAMP-generating agonists is not unique to the avian nasal gland. Similar effects have been recorded in rat parotid and cat submandibular salivary

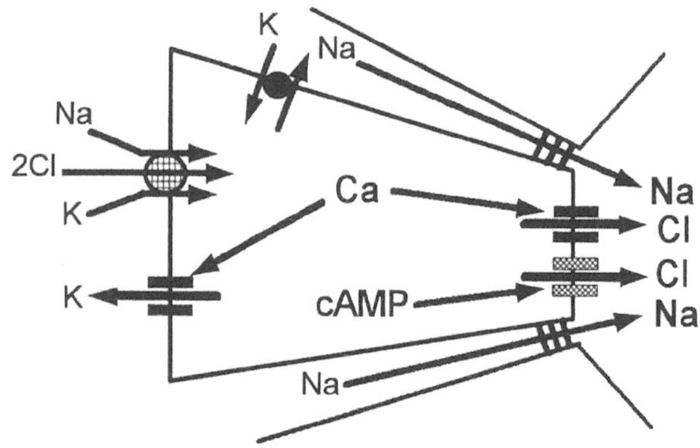

Fig. 12.7. Diagram illustrating the immediate sites of action of agonist-induced increases in intracellular Ca^{2+} and cAMP in the secretory mechanism in the avian nasal gland cells, and the interactions between these two signalling systems.

glands, as well as mouse lacrimal glands (Lundberg et al. 1982; Larsson and Olgart 1989; Larsson et al. 1990; Sasaki and Gallacher 1992). In these tissues, the cellular bases for such potentiation effects are generally unknown. Clearly, the mechanisms we have described above may well be have a more general relevance in the control of exocrine fluid secretion.

12.7 Conclusions

As stated at the outset, salt secretion in the avian nasal gland is primarily under cholinergic control via the release of acetylcholine from parasympathetic nerve endings, acting at muscarinic receptors to induce an increase in $[Ca^{2+}]_i$. These increases in $[Ca^{2+}]_i$ activate both apical Cl^- channels and basolateral K^+ channels to stimulate secretion. However, there is increasing evidence from a wide variety of tissues that, at physiological levels, agonist-induced $[Ca^{2+}]_i$ responses take the form of repeated oscillations, as we have indeed shown in the avian nasal gland. Our data indicate that such responses are only capable of a submaximal stimulation of secretory activity. We have shown that combinations of such low concentrations of Ca^{2+}-mobilizing agonists with agonists that act intracellularly to increase cAMP levels results in a marked potentiation of secretory activity to achieve essentially maximal effects, and we have suggested some of the cellular mechanisms that may underly such actions. The physiological relevance of these effects is supported by the finding of VIP-containing vesicles in the parasympathetic nerve endings in the avian nasal gland (Gerstberger 1988), suggesting possible co-release with acetylcholine. The gland is also known to receive a sympathetic innervation, elements of which parallel those of the parasympathetic system (Hasse and Fourman 1970; Gerstberger and Gray 1993).

Furthermore, clearly both VIP receptors and β-adrenergic receptors, both coupled to the activation of adenylyl cylase, are present on the secretory cells. The hypothesis we have presented describing how their actions on relevant membrane channels may interact with those of $[Ca^{2+}]_i$-mobilizing agonists such as acetylcholine and/or ATP to produce a potentiated response, suggest that such effects are likely to be of considerable significance to the overall control of the cellular activity of the gland in vivo and, perhaps, indicate interactions of relevance to the functioning of exocrine glands in general.

Acknowledgments. Research from the authors laboratory was supported by NIH grant GM-40457.

References

Berridge MJ (1990) Calcium oscillators. J Biol Chem 265: 9583-9586.

Berridge MJ (1993) Inositol trisphosphate and calcium signalling. Nature 361: 315-325.

Berridge MJ (1994) Relationship between latency and period for 5-hydroxytryptamine-induced membrane responses in the *Calliphora* salivary gland. Biochem J 302: 545-550.

Berridge MJ (1995) Capacitative calcium entry. Biochem. J 312: 1-11.

Bezprozvanny I, Ehrlich BE (1995) The inositol 1,4,5-trisphosphate ($InsP_3$) receptor. J Memb Biol 145: 205-216.

Bezprozvanny I, Watras J, Ehrlich BE (1991) Bell-shaped calcium-response curves of Ins(1,4,5)P_3- and calcium-gated channels from endoplasmic reticulum of cerebellum. Nature 351: 751-754.

De Young GW, Keizer J (1992) A single-pool IP_3-receptor-based model for agonist stimulated Ca^{2+} oscillations. Proc Natl Acad Sci USA 89: 9895-9899

Fänge R, Schmidt-Nielsen K, Robinson M (1958) Control of secretion from the avian salt gland. Am J Physiol 195: 321-326.

Finch EA, Turner TJ, Goldin SM (1991) Calcium as a coagonist of inositol 1,4,5-trisphosphate-induced calcium release. Science 252: 443-446.

Frizzell RA, Field M, Schultz SG (1979) Sodium-coupled chloride transport by epithelial tissues. Am J Physiol 236: F1-F8.

Gerstberger R (1988) Functional vasoactive intestinal polypeptide (VIP)-system in salt glands of the Pekin duck. Cell Tissue Res 252: 39-48.

Gerstberger R, Gray A (1993) Fine structure, innervation and functional control of avian salt glands. Int Rev Cytol 144: 129-215.

Hasse P, Fourman J (1970) The autonomic innervation of the avian salt gland. J Anat 107: 382-383.

Hildebrandt J-P, Shuttleworth TJ (1993) A G_q-type G protein couples muscarinic receptors to inositol phosphate and calcium signaling in exocrine cells from the avian salt gland. J Memb Biol 133: 183-190.

Hokin LE, Hokin MR (1964) Interconversions of phosphatidylinositol and phosphatidic acid involved in the response to acetylcholine in the salt gland. In: Dawson RMC, Rhodes DN (eds) Metabolism and Physiological Significance of Lipids. John Wiley Publishers, pp. 423-434.

Hoth M, Penner R (1992) Depletion of intracellular calcium stores activates a calcium current in mast cells. Nature 355: 353-356.

Hoth M, Penner R (1993) Calcium release-activated calcium current (I_{CRAC}) in rat mast cells. J Physiol 465: 359-386.

Iino M (1990) Biphasic Ca^{2+} dependence of inositol 1,4,5-trisphosphate-induced Ca release in smooth muscle cells of the guinea pig taenia caeci. J Gen Physiol 95: 1103-1122.

Larsson O, Olgart L (1989) The enhancement of carbachol-induced salivary secretion by VIP and CGRP in rat parotid gland is mimicked by forskolin. Acta Physiol Scand 137: 231-236.

Larsson O, Detsch T, Fredholm BB (1990) VIP and forskolin enhance carbachol-induced K^+ efflux from rat salivary gland fragments by a Ca^{2+}-sensitive mechanism. Am J Physiol 259: C904-C910.

Lowy RJ, Ernst SA (1987) β-adrenergic stimulation of ion transport in primary cultures of avian salt glands. Am J Physiol 252: C670-C676.

Lowy RJ, Dawson DC, Ernst SA (1985) Primary culture of duck salt gland II. Neurohormonal stimulation of active transport. Am J Physiol 249: C41-C47.

Lowy RJ, Schreiber, Ernst SA (1987) Vasoactive intestinal peptide stimulates ion transport in avian salt gland. Am J Physiol 253: R801-R808.

Lundberg JM, Ånggård A, Fahrenkrug J (1982) Complementary role of vasoactive intestinal polypeptide (VIP) and acetylcholine for cat submandibular gland blood flow and secretion. Acta Physiol Scand 114: 329-337.

Lytle C, Forbush B III (1996) Regulatory phosphorylation of the secretory Na-K-Cl cotransporter: modulation by cytoplasmic Cl. Am J Physiol 270: C437-C448.

Martin SC, Shuttleworth TJ (1994a) Muscarinic-receptor activation stimulates oscillations in K^+ and Cl^- currents which are acutely dependent on extracellular Ca^{2+} in avian salt gland cells. Pflügers Archiv 426: 231-238.

Martin SC, Shuttleworth TJ (1994b) Ca^{2+} influx drives agonist-activated $[Ca^{2+}]_i$ oscillations in an exocrine cell. FEBS Lett 352: 32-36.

Martin SC, Shuttleworth TJ (1994c) Vasoactive intestinal peptide stimulates a cAMP-mediated Cl^- current in avian salt gland cells. Regul Peptides 52: 205-214.

Martin SC, Shuttleworth TJ (1995) Activation by ATP of a P_{2U} 'nucleotide' receptor in an exocrine cell. Brit J Pharmacol 115: 321-329.

Martin SC, Thompson JL, Shuttleworth TJ (1994) Potentiation of Ca^{2+}-activated secretory activity by a cAMP-mediated mechanism in avian salt gland cells. Am J Physiol 267: C255-C265.

Michell RH (1975) Inositol phospholipids and cell surface receptor function. Biochim Biophys Acta 415: 81-147

Mikoshiba K (1993) Inositol 1,4,5-trisphosphate receptor. Trends Pharm Sci 14: 86-89.

Peaker M, Linzell JL (1975) Salt glands in birds and reptiles. Cambridge Univ Press, London.

Petersen OH (1992) Stimulus-secretion coupling: cytoplasmic calcium signals and the control of ion channels in exocrine acinar cells. J Physiol 448: 1-51.

Putney JW Jr (1986) A model for receptor-regulated calcium entry. Cell Calcium 7: 1-12.

Putney JW Jr (1990) Capacitative calcium entry revisited. Cell Calcium 11: 611-624.

Richards NW, Lowy RJ, Ernst SA, Dawson DC (1989) Two K^+ channel types, muscarinic agonist-activated and inwardly rectifying, in a Cl^- secretory epithelium: the avian salt gland. J Gen Physiol 93: 1171-1194.

Sasaki T, Gallacher DV (1992) The ATP-induced inward current in mouse lacrimal acinar cells is potentiated by isoprenaline and GTP. J Physiol 447: 103-118.

Shuttleworth TJ (1982) Amphotericin B and the elasmobranch rectal gland - implications for the relationship between oxygen consumption and ion transport. J Exp Zool 221: 255-258.

Shuttleworth, T.J. and J.L. Thompson (1987) Secretory activity in salt glands of birds and turtles: stimulation via cyclic AMP. Am J Physiol 252: R428-R432

Shuttleworth, T.J. and J.L. Thompson (1989) Intracellular [Ca^{2+}] and inositol phosphates in avian nasal gland cells. Am J Physiol 257: C1020-C1029.

Shuttleworth TJ (1994) $InsP_3$ receptor and intracellular Ca^{2+} release. In: Peracchia C (ed) Handbook of Membrane Channels: Molecular and Cellular Physiology. Academic Press, pp 495-509.

Shuttleworth TJ (1995) Intracellular signals controlling ionic and acid-base regulation in avian nasal gland cells. In: Heisler N (ed) Mechanisms of Systemic Regulation: Acid-base Regulation, Ion transfer and Metabolism, Advances in Comparative and Environmental Physiology, vol 22. Springer Verlag, Berlin. pp. 185-206.

Shuttleworth TJ, Thompson JL (1996a) Ca^{2+} entry modulates oscillation frequency by triggering Ca^{2+} release. Biochem J 313: 815-819.

Shuttleworth TJ, Thompson JL (1996b) Evidence for a non-capacitative Ca^{2+} entry during $[Ca^{2+}]_i$ oscillations. Biochem J 316: 819-824.

Streb H, Irvine RF, Berridge MJ, Schulz I (1983) Release of Ca^{2+} from a nonmitochondrial intracellular store in pancreatic acinar cells by inositol-1,4,5-trisphosphate. Nature 306: 67-69.

Taylor CW, Traynor D (1995) Calcium and inositol trisphosphate receptors. J Memb Biol 145 109-118.

Torchia J, Qu Y, Francis J, Pon DJ, Sen AK (1991) Carbachol-stimulated phosphorylation of a 170-kDa endogenous protein in avian salt gland cells. Am J Physiol 261: C543-C549.

Thorn P (1995) Ca^{2+} influx during agonist and Ins(2,4,5)P_3-evoked Ca^{2+} oscillations in HeLa epithelial cells. J Physiol 482: 275-281.

Wu JV, Shuttleworth TJ, Stampe P (1996) Grouped calcium titration curves: gating components of the calcium-activated K^+ channels may be heterotetramous. Biophys J 70: A192.

Zweifach A, Lewis RS (1995a) Rapid inactivation of depletion-activated calcium current (I_{CRAC}) due to local calcium feedback. J Gen Physiol 105: 209-226.
Zweifach A, Lewis RS (1995b) Slow calcium-dependent inactivation of depletion-activated calcium current. J Biol Chem 270: 14445-14451.

INDEX

MIX
Papier aus verantwortungsvollen Quellen
Paper from responsible sources
FSC® C105338

If you have any concerns about our products,
you can contact us on
ProductSafety@springernature.com

In case Publisher is established outside the EU,
the EU authorized representative is:
Springer Nature Customer Service Center GmbH
Europaplatz 3, 69115 Heidelberg, Germany

Printed by Libri Plureos GmbH
in Hamburg, Germany